For Barbara and Barbara Lee, for Margaret and Peggy, who thought I had something to say, and encouraged me to say it.

Low Oregon Grape
Berberis nervosa

Familiar

Friends

Northwest Plants

BY

Rhoda Whittlesey

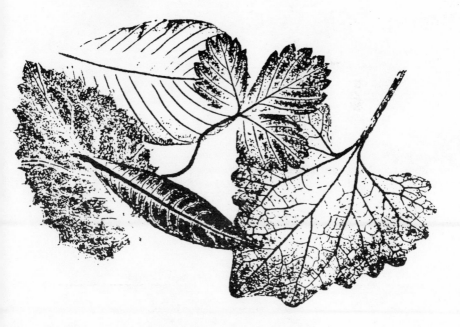

ROSE PRESS, PORTLAND, OREGON

ISBN No. 0–9615248–0–4

Cover by Karen Luse
Typeface: Benguiat
Printed by Raven Press, Lake Oswego, Oregon

ACKNOWLEDGEMENTS

Training with Camp Fire and with Tryon Creek Park for service as a nature guide helped me share my love of the out of doors. Experience as Nature Lady and tour guide made me aware of the need to consult many sources to learn uses and interesting facts which might increase others' enjoyment of local wild plants. Publishing this information appeared appropriate; all that research for my use alone seemed so wasteful!

Without the patience, understanding, and support of my husband, Charles Whittlesey, this book could never have been completed. My daughter Barbara Whittlesey and fellow Tryon Creek guide Margaret Harrang have reviewed, edited, and suggested through many versions, providing encouragement and prodding, as needed. Barbara Lee Orloff, first Volunteer Coordinator at Tryon Creek, and Peggy Robinson, another guide, who wrote *Profiles of Northwest Plants,* have been very helpful.

Sincere thanks are due to Dr. Herbert Orange, Head of the Horticulture Department of Clark Community College, who reviewed the contents prior to publication. The staff at Tryon Creek State Park, Tom and Marcia Pry of *The Bee,* Cathy Clark and Bill Helbock of The Raven Press, Katherine McCanna, and Jeffrey Zauderer, have been especially helpful, but it is not possible to acknowledge the tremendous support given me by many friends. Some of them may recognize their influence by text references; many have added direction to the final content. I have tried to include a variety of concepts which challenged my own interests; I hope this is useful to others who have similar concerns.

—Rhoda Whittlesey

Bishop's Cap, Leafy Mitrewort
Mitella caulescens

CONTENTS

Starters. 9

Nature Lady Notes 13

Classification. 21

NAMES. 25

Plant Particulars. 29

PLANTS. 31

SPINOFF. 185

Families. 189

People 196

Books. 200

Glossary 203

Index of Plants 209

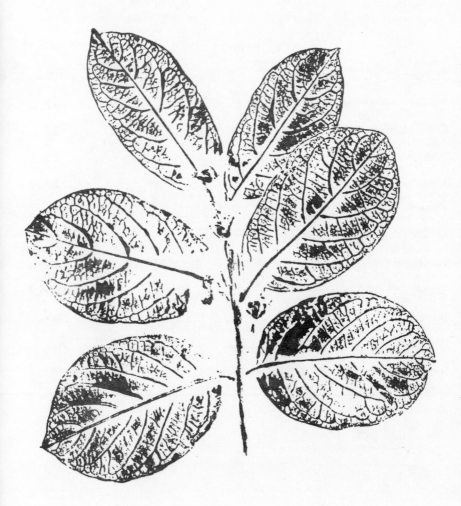

Twin-berry
Lonicera involucrata

STARTERS

Wild plants have always interested me, though my formal education included no botany. When I offered, close to twenty years ago, to share this enthusiasm with Portland area day campers, I was promptly assigned as Nature Lady. I find plants dependable; birds avoid nature walks with less-than-quiet kids. Hiking through camp, I learned how little I knew, and read a lot to earn the trust of campers who thought me An Authority. I noted on 5 by 8 cards what authors said about plants, and took a bucket with these notes and magnifying glasses on hikes; it became my Bucket of Brains.

My concern has been to relate plants to our lives. Youth group leaders, school teachers, park guides, all like to pique interest by adding sidelights with the comment "This is a Nettle; it will sting you!" No one source had all I wanted; I found a medicinal use here, a food or materials use there. When I started looking, few books told of local plants. Haskins' classic wildflower book, Hill's Spring Flowers of the Columbia River Valley, and Lyons' Flowers to know in Washington had more than identification, but these were not enough. I read anything I could find about local plants and shared uses for plants in the differing habitats of five Camp Fire day camps in Portland and Forest Grove. Since starting as a tour guide at Tryon Creek Park in 1977, I have added to my list.

Many books about wild plants have been published in recent years. The most exciting was by Clark, a fantastic photographer and scientist. His pictures really identify plants and his remarks about them are delightful and informative. Field guides based on his text and photographs are again available. I do like books with pictures!

The public library is wonderful; browsing near a book I have used gives more help. My own library has perhaps 50 books which I consult frequently and others I use less often. Many relate to limited fields; some emphasize identification, some tell of plants used for food, materials, or medicine; some are locally oriented, others about other areas. Many have been published since I started my card file. Older volumes include scientific names no longer used. Probably the most fascinating is Grieve's Herbal, which includes references for food, medicine, materials, literature, early herbalists, and classical scholars. Mrs. Grieve was a practicing herbalist during World War I. When I learned to find older plant names—synonyms—in the large Hitchcock botany, I found material missed in earlier passes through books written when those names were current.

This miscellaneous information, in a combination not found elsewhere, has challenged my interest. All plants have been checked in the same basic sources; uses, or lack of them, were confirmed in several references. Absence of a plant in a specialized source is a helpful reference. Those not found in poisonous plant books are probably not poisonous; those not in herbals may not have had herbal medicine use. Controversial plants have been researched extensively. I tend to evaluate wildfood books by reading about Red Elderberry; a statement that it is poisonous makes me distrust the research. Conflicting information is also published about Bittersweet Nightshade.

For plants with few comments, I found little but identification. Reasons for no reported use could be many: perhaps they were scarce; perhaps texture or flavor of nonpoisonous plants made food use undesirable. Seasonal availability limited ethnobotanist Erna Gunther in her research with Western Washington Indians; many more plants probably were used, or more widely used. The later publication by Nancy Turner of ethnobotany for Coastal and Interior British Columbia peoples is more extensive. Most Indian uses came from these books. Sweet told of California Indians, usually without group names; Mrs. Murphey, a longtime employee of the Indian Agency, worked with groups no longer living in their original locations. Her references were particularly helpful for plants east of the Cascades and in Nevada.

Because different groups had different uses, it seemed important to give group names where these were available. Indians in British Columbia included: Straits Salish, Halkomelem, Squamish, Sechelt, Comox, Bella Coola, Nootka, Southern Kwakiutl, Central Kwakiutl, Coast Tsimshian, Thompson, Lillooet, Okanagan, Shuswap, Kootenay, Niska, Gitksan, Tlingit, Carrier, Stali, Okanagan, Saanich, Squamish. Washington groups are: Lummi, Samish, Swinomish, Snohomish, Klallam, Makah, Quileute, Skokomish, Puyallup, Nisqually, Squaxin, Chehalis, Chinook, Cowlitz, Snuqualmie, Skykomish, Skagit, Upper Skagit, Muckleshoot, Green River, Lower Chinook. Oregon groups mentioned are Tillamook, Sandy River, Washoe, Molalla, Callapooya, Warm Springs, Klamath, Takelma, Willamette Valley. Presence or absence of groups in this list indicates available information, not that the coverage is complete, even for groups about which some ethnobotany has been published.

Traditional, herbal, and Indian medicine did not necessarily prescribe the same uses. Some plants had only herbal, some traditional, others only Indian medicinal use. I have tried to point out differing uses in the different disciplines. Frequent preference for freshly gathered plants increases the difficulty of using botanicals. Botanical medicines, difficult to preserve and subject to considerable variations, have been largely replaced by synthetic compounds, so much literature consulted reports former uses. I have no intention of collecting plants for medicine. I do not recommend collecting medicinal plants, an endeavor for which specialized knowledge is essential, nor do I suggest use of such medications without consulting appropriate professionals. I do not make a practice of harvesting and eating wild plants, but it interests me to know which have provided food.

Using leaf prints as illustrations was a result of enjoying these as a camp craft; prints have been reduced to fit space needs. A little about classifying and the mechanics of plant names lessens the fear of the unknown. Scientific names became more logical to me as I worked my way through "What's it good for." Knowing name sources has made me aware of distinctive features, a help in remembering which plant is which. Names become less intimidating as we see that they tie together plant characteristics, history, explorations, medicines, or individuals associated with them. I have identified some individuals known through history, some important in plant study or science. The glossary gives words foreign to our normal vocabularies. Family and plant listings with name sources, food, materials, medicinal and other uses provide an overview of how plants enrich our lives.

The more than 200 plants included were identified in the greater Portland area, but are typical of much of the Pacific Northwest. It is rewarding to find and compare them in different locations. Hopefully the leaf prints, name sources, and botanic terms will be useful, but this is not meant for an identification tool; there are now many for Northwest plants. This is a browsing book about familiar friends, a sharing of highlights which have helped and fascinated me.

Lemon Balm
Melissa officinalis

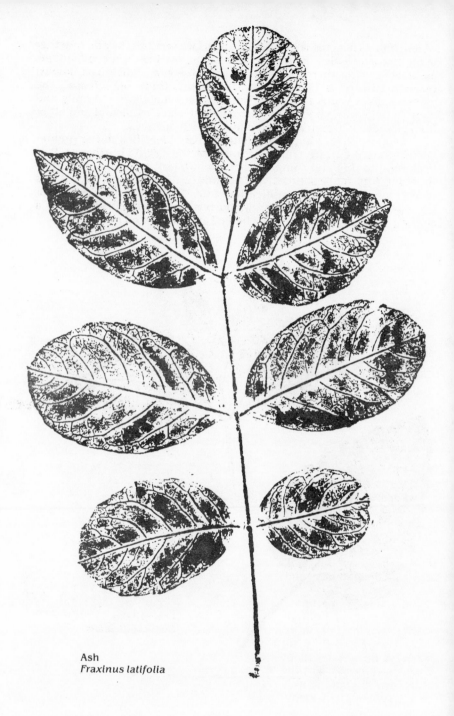

Ash
Fraxinus latifolia

NATURE LADY NOTES

Plant use precedes recorded time. Plant medicines were noted in Sumerian cuneiform tablets dating back to 4000 B. C. Little written information about early agrarian civilizations of Egypt, China, the Assyrians, pre-Aztec Indians, and others was available to European scholars who provided our heritage. In the Middle Ages science was largely in the hands of clergymen and friars; in the 16th century Protestant clergymen were also involved. Some writings of Cratevas, a Greek physician of the first century B. C., and those of the Romans Dioscorides and Pliny the Elder of the first century A. D., were preserved in monasteries, copied in scriptoria and made available after the Renaissance. This classical material, translated into many languages, was included in herbals, which were largely concerned with medicine. In the latter part of the 17th century botany began to separate from medicine.

Many less costly herbals made possible by Gutenberg's invention of movable type were widely distributed from 1470 to 1670. As study progressed, European botanists realized that local plants differed from Mediterranean species described by Dioscorides. The 1551 herbal of William Turner, written in English rather than Latin, was noteworthy because it listed some English plants! These brief comments suggest botanic history, but interest comes alive in the growing plants.

Opening eyes to outdoor wonders is a joy and a challenge. A color search is an eye opener; look for the many shades of green. What other colors can be found? Do these reflect changing seasons? Are colors equally visible to pollinating insects? Slashes or lines of color on flowers attract bees, which see the color range from blue to yellow, plus ultraviolet. Butterflies see yellow, orange, and especially red. Animal homes may not be obvious, but food is always evident: leaves, pollen, nectar, seeds, cone scales, hulls of nuts, beaver-cut trees, holes in trees. Were holes made by woodpeckers or crawly critters? Who ate at this cafeteria? Are food remnants in the spider web? The Northwest has many harmless spiders which make lovely webs; these are best seen in the fall.

Try looking at plants closely, perhaps with a magnifying glass. Are stems round or square, fibrous or juicy, smooth or hairy, dry or hard; are leaf edges smooth, notched, or perhaps deeply cut; are leaves small or large, thick or thin, leaf veins parallel, palmate, pinnate? Is leaf arrangement opposite, alternate, spiral, or whorled? Are leaves at the plant base shaped differently than those on the flowering stem? Are flowers symmetrical or asymmetrical, one or more colors? Are flowers flat or belled, a long tube or arranged in heads to attract insect visitors? Do blooms open first at the top or bottom of a tall stalk? Do they hang down or lift their faces up? Are flower stems long or short, with or without leaves, or are flowers without stems? These characteristics help identify plants.

Star-flowered False Solomon's Seal
Smilacina racemosa
parallel veins, leaves alternate on stalk, smooth
leaf edges; end of flowering stalk

Look for springboard notches on stumps left by early loggers.
These notches were precisely cut to fit an iron piece bolted to the end
of a board which supported the timber faller; trees were cut above the
swell and above rotten wood often found at the tree base. In Tryon
Creek Park notches still visible from logging about 1900 give an idea of

Duckfoot
Vancouveria hexandra
opposite leaves, palmate veins; end of leaf stalk

how long it takes there to break down wood for recycling into soil. Size of trees near stumps or burned snags also shows time elapsed since logging or forest fires. Such things are obvious, but how do they fit together? Do variations serve a purpose? In what ways?

Plants, part of a complex living community, are particular where they live. Some like shade, others prefer sun, some like the mountains, some the seashore, some flourish where there is much moisture; others have learned to live in dry areas, yet how many seem to grow anywhere! We may consider them weeds: plants growing where they are not wanted. The same plant elsewhere could be welcome. An ecosystem (eco from the Greek oikos, house), consists of the living community plus nonliving environmental factors such as moisture, soil, light. Ecology tries to explain the balance between plants and their surroundings. It also deals with manipulation of plants to improve life. Selected seeds and cultivation were used to alter edible plants before the time of written records. Reforestation, extensively developed in this area only within the past fifty years, is one effort to improve life through controlled planting.

Geography tells of plant migration routes and physical barriers to plant distribution. Geology reveals past climates and changing land formations. Fossil records confirm that similar plants have lived together for millions of years, changing and developing, evolving, to fit their surroundings and take advantage of available food, moisture, light, and shade. Each species now living has achieved a balance within its habitat. Those unable to compete or change with changing conditions are now extinct. Man's activities have greatly affected these changes.

Plant habitats have been divided for convenience into what are called life zones, based largely upon distance from the equator. Mountains, rivers, temperature, wind conditions, available moisture, etc., affect growth and distribution of plants. In general, the farther north, or the higher the elevation, the colder are the growing conditions. Life zones help identify similar habitats or zones where certain plants usually grow. Boreal zones are northern in aspect, austral, southern. Also descriptively named is the Arctic Alpine zone; those called Hudsonian, Canadian, and Transitional are less obvious. For a large picture, this concept provides a good overview; like all generalizations, it is subject to many exceptions, one type of which is mentioned next.

Areas called botanic islands, which contain plants not typical of similar areas nearby, evolved through accidents of geology, geography, or changing climatic conditions. Saddle Mountain has cold-climate plants which were able to adapt when the weather warmed appreciably, and was high enough to become a refuge for some plants eliminated nearby when the weather became too warm for their normal life patterns. Some of these now grow no closer than the Columbia Gorge, another botanic island, or the Olympic mountains of Washington, an expected habitat: cool areas farther north or higher in elevation. Many plants now on Saddle Mountain thus remain as living examples of what is in some part an earlier vegetation in that area. Mary's Peak near Corvallis and the Camassia Wilderness near West Linn are other botanic islands. We can enjoy in the varied flora of these "islands" plants now more commonly found in other life zones or habitats.

Plants produce and reflect change. Living and decaying vegetation help control water, providing protection from excessive rain and snow which could be disastrous. As water plants die and decay, they build up layers which change a pond to a swamp and eventually to solid ground. As tree growth reduces light on the forest floor, low-growing plants may find it difficult to survive. If they react by failing to set seeds, they may not reproduce and will be replaced by plants able to live with less sun and light, so undergrowth gradually changes as forests grow older.

Basic to the existence of all living things is photosynthesis, the process by which green (chlorophyll-bearing) plant cells manufacture sugar from carbon dioxide and water in the presence of sunlight, with oxygen as a by-product. Differently stated, photosynthesis allows plants to store energy by changing light energy to chemical energy. Non-green plants cannot manufacture food and must obtain it from other organisms. Plants are the food base for other life forms. Primitive or simple plants which have survived with little change are algae (water plants), mosses, seaweeds, bacteria, lichens, and fungi; these continue to affect our lives extensively.

Fungi are non-green, taking their food from other forms, and are agents of decay, making organic material available for reuse, often when we wish they wouldn't. Some of these are molds, yeasts, mildew, and rusts, which cause much plant disease, and mushrooms, which we may consider treats. Mushrooms have two ways of releasing spores: some have "gills" on the under surface, and the spores drop down in the pattern of the gills. The "pore" type of mushroom has an apparently closed under surface, dotted with tiny openings, through which the spores fall when they are ripe.

Lichens, (Greek *leichen)*, thallus plants growing on rocks and trees, are composed of algae, (singular, alga), and fungi, (fungus), growing together in an association called symbiosis (Greek *sym*, with, *biosis*, to live). The fungi supply moisture and shelter, the algae food manufacturing ability through photosynthesis; other nutrients come from moist air and rain. Some scientists now suggest this relationship may be parasitic rather than mutually beneficial. Lichens grow slowly, none reported to grow more than an inch and a half a year, most more slowly. Some are believed to live four hundred years, some up to as much as four thousand.

Broadly speaking, there are three types of Lichens: crustlike, crustose; leaflike, foliose; and bushlike or shrubby, fruticose. Some look like mosses. Pollution can be monitored very closely by presence or absence of lichens, which are more vulnerable than any other type of plant. Primitive plants with no vascular system, lichens absorb soluble elements directly through the plant surface. They have no way to cleanse the system through pores (stomata), or by dropping leaves, as do more complex plants. (Pine trees in heavily polluted areas will drop their leaves as do deciduous trees.) Elements not useful to the plant interfere with life processes and result in its death. A major pollutant is sulfur dioxide, present in smoke and industrial discharges. Experts can determine degree of pollution not only by types of lichens, but even by quantities of individual species. The fruticose are the first to go; the crustose—with possibly less surface exposed for absorption—are most durable. Lichens have been used to make litmus paper, used to test for acid or alkali content. They have served as fixatives in perfume; as a demulcent, emollient, and emulsifier, as a folk remedy for bronchitis. Use of lichens as natural dyes is no longer considered ecologically desirable, because of the long time required to grow the large quantities necessary for dyeing.

Mosses are prolific and live in moist shady locations, often where growing conditions are unsuitable for other plants. Mosses do not have true roots or true leaves, and have no circulatory system which would permit their growing to greater size. Acids released by decomposing mosses help break down rocks upon which they may grow, thus building up soil. This new soil makes it possible for other plants to become established. Ferns and Horsetail were a dominant part of vegetation in Paleozoic times, over 200 million years ago, when the land was wetter. Their reproduction involves two alternating life forms and requires water to bring together male and female parts for fertilization.

Evolutionary changes made it possible for plants to flourish and reproduce in dry-land conditions. Developing the ability to bring together male and female reproductive parts without the presence of water was an important advance for land plants, first appearing in the Gingko tree. When plants developed a single type to replace the two life forms of the spore-bearing plants, resources were used more efficiently. Energy saved enabled seed plants to become the dominant form of vegetation; an estimated quarter of a million species in more than 300 families are now known.

Flower color, pollen, nectar, shape, size, and fragrance all developed to help fertilization by bees, ants or other insects, birds, the wind, or gravity. Flower shape may help push pollen off on bodies of insects, which then transfer some pollen to the next flower visited. Complex adaptation for insect pollination indicates long association between insects and plants. Brown and purple flowers are pollinated by insects; low-growing flowers may get help from earthbound crawlers. Inconspicuous wind-pollinated plants such as grasses and Ragweed are considered advanced in evolutionary development; they produce great quantities of light, dry pollen, usually lack odor and/or nectar, and are not attractive to insects.

Seeds store food efficiently to help plants grow when conditions permit; seeds 1,000 years old have been known to germinate. Seeds fly, float, hitchhike, and attract feeding animals with the help of structural adaptations designed to spread them. Some pods snap open, propelling seeds away from the parent plant. Winged and plumed seeds drift or soar with the wind. Many seeds pass undamaged through digestive systems of animals which ate the fruit. Flowers precede fruits, a broad term botanists use not only for juicy structures but also for vegetables and grains—corn, wheat, oats. Nuts, berries, grains—all seed forms or carriers—provide a major part of our food supply and that of animals sharing our space. Some plant techniques assure seed production without outside help; Violets have small, closed self-pollinating flowers in addition to the open ones. Seeds provide the most widespread method of plant reproduction, but many plants also multiply by vegetative means such as root divisions, bulblets, etc.

Age, light, water and nutrition affect plant shapes. Under differing conditions leaves of the same species may be thinner, broader or smaller, stems taller, shorter or fatter (storing more water). Reproductive parts are unique in each species; these flower and fruit parts are less involved in the outside environment, hence less likely than are leaves and stems to change due to variable growing conditions. This makes them more useful in tracing relationships by comparison with fossils and primitive living plants. Relatively little plant material survived in fossil form, so it is helpful that the distinctive pollen grains often reached wet and swampy areas and were preserved through geologic time.

Many forest plants bloom before deciduous trees develop leaves, taking advantage of sunlight available then. Have you noticed how flower stems stretch above leafy parts, bringing seeds higher for better distribution? Although we may see this, we may not recognize its importance, just as we often do not realize how people influence and affect life patterns of plants and animals.

Plants absorb more water than they use; water is returned to the atmosphere by transpiration (Latin *trans*, through, plus *spirare*, to breath). Leaf openings called stomata return unneeded moisture to the air; this process also serves as a cooling technique. Under sides of leaves often have more stomata; moisture loss there is better controlled as they are less exposed to severe or abnormal weather conditions. Somewhat comparable openings in tree bark, lenticels, permit exchange of gasses between outside air and the living portion of the tree under the bark. The name, from the Latin *lenticella*, a small lens, refers to the pore shape. Lenticels are especially visible in the bark of Cherry trees. Bark protects inner growth from rapid temperature changes and from fire. Some trees grow new bark each year; some keep the same bark, which gets heavier and thicker each year, adding insulation. Bark often splits vertically as it grows, adapting to increased trunk size.

Drying winds and freezing of soil moisture have the same effect as drought, because they deprive plant systems of moisture. As cool winter weather slows life processes, trees need to reduce moisture loss. Deciduous trees developed the protective device of dropping leaves, the part which loses the most moisture. The joint between leaf and tree is sealed with the removal of the leaf so the plant is not open to damage. Strong evergreen leaves serve for more than one growing season. Leaves of cone-bearing evergreen trees are adapted to needles; the lesser leaf area conserves moisture when leaves are not actively producing food, and provides a surface less vulnerable to winter damage. Needles last from three to eight years, depending on the species.

Manufacturing food is a basic function of leaves, which are sometimes modified for food storage and plant protection. Leaves of succulents (Latin *sucus*, juice), have special water storage tissues in

19

stems or leaves; some are also waxy, which discourages evaporation and helps survival in dry weather. Hairy leaves are protective as they retard evaporation, provide shade from light and heat, and foil or frustrate predatory insects. Leaves of herbs in wooded areas tend to be larger than those of the same species in more exposed areas, the stems longer and less strong, as these plants are less susceptible to damage from strong wind and drying sun and need to take advantage of all possible light and sun. Mesophytes are plants growing under medium to normal moisture conditions (Greek *mesos,* middle, *phyton,* plant); xerophytes are resistant to drought or adapted to growth in very dry places, (Greek, *xeros,* dry). A thick cuticle, reduction of transpiration surface, water storage tissue, and the ability to close stomata are adaptations which help survival when water supply is limited.

Stems give support, conduct food, water, and inorganic salts, produce new growth, or may be adapted for photosynthesis, food storage, or protection. Spines and thorns may be modified stems. Root systems provide anchorage, storage, conduction and absorption of water and minerals. Biennial and perennial plants often use this storage to produce flowers after the winter resting period, before the plant resumes manufacturing food.

Tree roots give support and provide water and minerals which circulate to supply life to the whole tree. Roots of herbs often penetrate deeper into the ground than those of many trees. When there is plenty of moisture near the surface, tree roots may be shallow. Trees in the center of a dense stand, protected from winds, may also have shallow root systems and be susceptible to blowdown when exposed to heavy wind.

Cross sections of trees growing in temperate zones show a light-colored ring of rapid spring growth and a darker, usually narrower, ring of summer growth. Age of trees can be estimated by counting these annual rings or by counting whorls of branches on conifers, which produce one whorl per year.

Candy Flower, Siberian Miner's Lettuce
Montia sibirica

20

CLASSIFICATION

People have always liked to have names for things and to arrange or classify them in a somewhat orderly fashion. Botanists combine plants in groups called families, genera, and species. (To simplify discussion here, taxa larger than the family have been omitted.) A family has several genera (singular, genus, from the Greek *genos*, race or stock). Each genus has one or more somewhat similar species, (plural, also species). A species identifies a single plant type, a group of similar individuals able to reproduce. These terms make it possible to identify an individual plant with two words. See also chapter on plant names, taxonomy and phylogenetic in Glossary.

Plants without seeds, such as algae, mosses, liverworts, clubmosses and fungi are considered more primitive, as they do not have true roots, stems, leaves, or circulatory systems. All reproduce by spores and their life processes involve two distinct types of plants, one producing spores, the other, sex organs.

Liquids circulate throughout more complex plants in what is called a vascular system, (Latin, *vasculum*, diminutive of *vas*, vessel), which helps maintain life farther from the surface, hence permits plants to grow taller and/or larger. Plants with vascular systems have true roots, true stems, true leaves. Ferns and Horsetail reproduce by spores and are called "lower" vascular plants; "higher" plants reproduce by seeds.

The two divisions of seed plants, gymnosperms and angiosperms, show progressive development. Gymnosperms, most conspicuous of which are the conifers, are considered less advanced, because the seeds are "naked"; (Greek *gymnos*, naked, *sperma*, seed). At the time of pollenization, the ovule is freely exposed and the wind-distributed microspore is deposited on it. Gymnosperms do not have flowers. Angiosperms, flowering plants, have seeds enclosed within and receiving protection from the ovary; (Greek *angeion*, a vessel); pollen is deposited upon a specialized plant part, the stigma. Flowering plants have a wide variety of seed dispersal techniques. Color, shape, scent, nectar and pollen are all tools which attract pollinating insects and animals.

Many new species and genera were discovered after the Renaissance, as scientists accompanying explorers collected seeds and specimens of newly found plants. Increasing numbers of known plants intensified the need for systematic classification.

The first generally-accepted system of plant classification was published in 1753 by the Swedish naturalist Carolus Linnaeus. He coordinated and combined work of earlier students. Many names he established were known to his contemporaries; some had been used since ancient Roman times; some are still in use. Linnaeus' great contribution was establishing the binomial system, which makes it possible to identify individual plants with only two words or parts, the genus and the species. These are the scientific names of plants; they have remained relatively stable.

Classification attempts to reflect order of development, placing primitive first and highly-developed groups later. By the end of the 19th century botanists agreed in general about classification of plants less advanced than the flowering plants. Botanists continue study of fossil materials and existing plants, trying to decide how flowering plants developed by identifying like features. Most families include plants with both primitive and advanced characteristics. Linnaeus identified approximately 6,000 plant species in 1,000 genera. Present-day scientists have identified more than 200,000 species, world-wide, of flowering plants alone, without counting lower vascular and more primitive plants. This suggests some of the problems and progress of systematic botanists.

Early identification systems were based upon convenient and readily-observed features; for flowering plants Linnaeus used stamens and pistils—the so-called sexual organs of a plant. Such systems are considered artificial as they do not include all traits; the goal is to identify natural relationships. The larger groupings established by Linnaeus have undergone many changes—as he anticipated would occur as additional knowledge became available.

Important botanists in the late 19th and early 20th century developed and presented phylogenetic systems which were widely accepted and used. The American teacher and botanist Charles Edwin Bessey was the author of one system; his work had profound and continuing influence. The German botanist Adolf Engler published a system for classifying flowering plants which was widely used. Each arranged plant orders and families in what was considered a more natural sequence than those of earlier systems. These were basic standards for many years, changing to include new information. The two major groupings of flowering plants are monocotyledons and dicotyledons, having one or two seed leaves. (Greek, *monos*, solitary, *dis*, twice, plus *cotyledon*, a cup-shaped hollow.) Engler and Bessey regarded monocots as more primitive than dicots.

One way classification enters our experience is in the sequence of plants in wildflower books and guides. Families are presented in the order considered natural, rather than alphabetically. Monocyledons are in one sequence, dicotyledons in another, all listed in phylogenetic order. Guides arranged by color follow a similar sequence under each color. This conventional manner of arrangement can be confusing.

Changed interpretation of available knowledge, added information and scientific techniques not available earlier resulted in a need for reevaluation of sequence of plant orders. Systematic botany studies published in the late 1960s show major changes. Among those publishing such works are the American Arthur Cronquist and the Russian Armen Takhtajan, who share many common concepts. One difference reflects a change in sequence of monocots and dicots. Absence, loss, or discontinuation of a plant feature is considered progress, as it reduces the amount of energy required to support life processes. Systematic botanists now believe that dicotyledons lost one seed-leaf in the evolutionary process, making monocots (with only one

22

seed-leaf), a more advanced form. In these classifications monocots—Lilies, Grasses, Orchids—are presented last. College textbooks, one of which is Hitchcock & Cronquist's Flora of the Pacific Northwest, are beginning to use this sequence.

A technical sequence of flowering plants would start with the plant kingdom; the next smaller taxon is a phylum; (flowering plants are considered a phylum). As presented by Cronquist, within the flowering plants—angiosperms—are two classes, Magnoliatae, (dicotyledons), and Liliatae, (monocotyledons); ten subclasses, six of which are in Magnoliatae, 4 in Liliatae. Next smaller taxon is an order, names of which end with -ale. Endings of families, next smaller taxon, are suggested as -aceae; this is the largest taxon with which comments here are concerned. Genus and species are considered at greater length in the chapter on names.

Mountain Ash
Sorbus aucuparia

Tall Oregon Grape
Berberis aquifolium
leaf with nine leaflets

NAMES

Plant names are confusing, as both individual and regional use varies greatly. You may use any common name you wish and not be wrong, but this may not tell someone else what plant you have in mind. Almost every yellow Violet is called Johnny-jump-up by someone. If it makes a difference to you whether this is a Wood Violet, a Prairie Violet, or perhaps the Evergreen Violet, you may want to use a more descriptive name or learn the scientific name.

Some familiar names have Anglo-Saxon roots. Angles and Saxons were Germanic tribes which conquered southern and eastern England and southern Scotland in the fifth and sixth centuries. *Wort*, or *wyrt*, the Anglo-Saxon, Old English or Middle English word for plant, is often used now in combinations: St. Johnswort, Coolwort, Figwort, Toothwort.

A recent question about the English name of *Montia sibirica* sent me to 35 wildflower books, and confirmed that "common" names are far from commonly used. Results: Candy Flower, 7; Siberian Miner's Lettuce, 5; Western Spring Beauty, 8; Spring Beauty, 3; Siberian Candy Flower, Siberian Lettuce, Siberian Montia, Indian Lettuce, one each. Not all included this species, nor did all give common names.

It seems to me a name should say something about the plant. Candy Flower has meaning: delicate stripes on the blossoms vaguely resemble striped candy. Siberian Miner's lettuce—a cumbersome name—says two things: this plant grows in Siberia; miners ate it. The name Western Spring Beauty suggests that someone remembered a flower from "back home" and added Western. Other quite different plants are also called Spring Beauty. One, *Cardamine pulcherrima*, blooms in late February or early March and is almost gone before *M. sibirica* appears a month or so later.

Names including Western or False reflect choice of familiar names for local plants like those settlers knew elsewhere. Western Spring Beauty, if you use this name for *M. sibirica*, and Western Wild Ginger are related to eastern plants. False Solomon's Seal is a different genus than Solomon's Seal, but both are in the Lily family. False Hellebore is in the Lily family, Hellebore, in the Buttercup family.

Scientific names are not really frightening; some are fun to roll off the tongue. Do you think twice when speaking of a Chrysanthemum, a Rhododendron, or a Trillium? These are genus names used as common names. They do not say which species of *Chrysanthemum*, *Rhododendron*, or *Trillium*, but you're in the ball park.

Scientific names are a tool for exact identification all over the world, as they differentiate one plant from all others. They have two parts and are called binomials, (*bi*, two, *nomen*, name). The first part is the genus, the second is the species. The genus name is used in only one family, never duplicated. This permits identification of an individual plant with two words. Species names may be reused in different genera. These often describe some feature of the plant or its habitat, as does the Latin *arvense*, of the field, in *Cirsium arvense*, Canada Thistle, and *Equisetum arvense*, Horsetail, both of which flourish in fields.

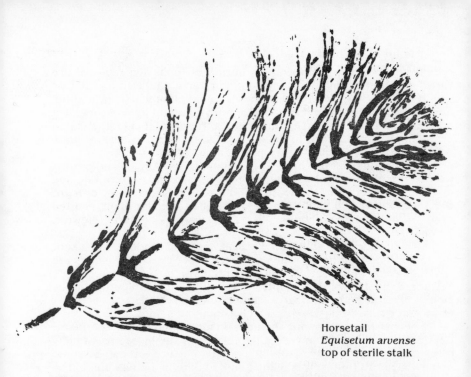

Horsetail
Equisetum arvense
top of sterile stalk

The genus name is capitalized; the species name is usually written with a lower case letter, even if it comes from a proper name. There is no required pattern of word endings for genus and species names.

The binomial system starts with names used by Linnaeus in 1753. For plants discovered later, the identifying scientist, called the authority, proposes a name. First publication, which includes filing a type sample with flowers, seeds, above-ground parts, root systems, etc., establishes a scientific name.

These names conform to the International Code of Botanical Nomenclature, introduced in 1867 and periodically updated. These mystifying mixtures of Latin or Greek words or people's names usually have some logic. Latin has long been the accepted scientific language, so combining Latin words to make plant names is logical. While Latin is not familiar to most people of our times, Latin roots are part of many words we use every day. Words from other sources are altered to fit Latin usage. Adding the suffix *-ensis* to a place name is such an alteration: *Cornus canadensis*, Canada dogwood; *Osmorhiza chilensis*, Chilean Sweet Cicely.

In Latin, word forms change for varying uses and sexes. Typical Latin nouns in the nominative case—those used as the subject of a sentence—end with *a, us,* or *um* to identify feminine, masculine, or neuter forms. Genus names, or species names which modify them, are usually in the nominative case. The genitive case, which translates as "of", is often used for a species name. One derived from a man's name may end with a double *i, douglasii, menziesii, nuttallii,* (unless the final

26

letter is an *r*, in which case only the single *i* is used, *hookeri*). For a woman's name, *iae* or *ae* is used, depending on the letter it follows. These are the Latin forms for the genitive case, so *Cornus nuttallii* is the Dogwood of Nuttall, or Nuttall's Dogwood. What we may consider peculiarities of Latin forms are not about to change, but if we understand some whys and wherefores it may be easier to use and spell such words.

Deciphering scientific names provides an interesting challenge: *Trillium* has parts in threes. *Equisetum*, combining Latin for horse, *equus*, and bristle, *seta*, pictures the Horsetail; the common name is a translation. *Folium* is Latin for leaf; *pinnatifolia* is a leaf cut into many parts or *pinnae*, like those of a fern. Sometimes Greek words are used: *macrophyllum*, a large leaf, (*macro*, large, *phyllon*, leaf). *Officinalis* as a species name reports medicinal use, recognized by Linnaeus in *Taraxacum officinale*, Dandelion, and *Melissa officinalis*, Lemon Balm.

An error in the first publication of a name will be perpetuated, as illustrated by the genus *Penstemon*. The Greek *pente*, five, and *stemon*, thread, were chosen for these plants, which have five stamens. A typographical error left out the first *t*, so *Penstemon* is the accepted spelling.

The logic of changing scientific names is not always apparent. Reclassification may be the result of study of new materials, using new equipment or techniques, additional study of previous materials, or perhaps discovery of an obscure earlier name which was properly published and hence entitled to acceptance. Plants difficult to classify have been changed from genus to genus, species to species, as botanists evaluated color, stature, shape of leaves, size of petals, type of root systems, details of reproductive systems, etc.

Names no longer accepted as correct are called synonyms; some are included with plant listings. Some synonyms were the names of choice when books now available as reprints were originally published.

Occasionally names are returned to an earlier usage. The Northwest Stinging Nettle was formerly called *Urtica lyallii*. When botanists decided that this was the same plant Linnaeus described as *U. dioica*, *U. lyallii* became a synonym. Some botanists add variety *lyallii* when referring to Northwest Nettles: *U. dioica*, var. *lyallii*. (Where several references are made to the same genus and the meaning is clear, it is customary to abbreviate the genus name after the first listing: *Urtica lyallii*, *U. dioica*.)

Those who disagree with the reasons for a change may continue to use the previously accepted name. A confusing example is Oregon Grape, identified as *Mahonia*, or *Berberis*. Pursh classified *Berberis aquifolium* and *B. nervosa* in 1814, using a genus published by Linnaeus. In 1818, Nuttall reclassified some *Berberis*, establishing a smaller new genus, *Mahonia*, which included mostly evergreen shrubs with compound pinnate leaves. Because respected authorities do not agree on which genus is correct, both *Berberis* and *Mahonia* continue in use. Hitchcock uses *Berberis;* Hortus Third, considered the present authority on scientific names of cultivated plants, accepts *Mahonia* as the name of choice. *Mahonia* is also sometimes used as a common name. Do not let the technical consideration of whether the correct name is that given by Pursh or by Nuttall deprive you of information. *Berberis* and *Mahonia* are used, by different individuals and in different books, for the same plants; references to *Berberis aquifolium* or *B. nervosa* apply also to *Mahonia aquifolium* or *M. nervosa.*

As the number of known plants is astronomical, limiting the number of species is helpful. Those who work with naming plants may be "lumpers" or "splitters". A "lumper" tends to put more plants in a single species, believing some differences, perhaps due to local environment, not sufficient to justify a separate species. Hitchcock, the authority used here, identifies Foamflower, *Tiarella trifoliata,* as a single species with three varieties: *T. trifoliata* var. *trifoliata; T. trifoliata* var. *unifoliata,* and *T. trifoliata* var. *laciniata.* A "splitter" may give greater weight to the same characteristics, and identify these as *T. trifoliata, T. unifoliata,* and *T. laciniata.*

A more complete binomial adds the name, usually abbreviated, of the individual who first classified the plant. *Galium aparine* L. says that Linnaeus described this Bedstraw. When a plant is reclassified, the name of the first authority is put in parentheses, and the author of the presently accepted classification is added. *Pteridium aquilinum* (L.) Kuhn says Linnaeus identified this fern and Kuhn reclassified it. The family chapter gives these names, which are informative though brief. It is exciting to note how many plants Linnaeus included, and how many honor early botanists or explorers such as Nuttall, Douglas, Pursh, and Hooker.

I like Canada Thistle: *Cirsium arvense* (L.) Scop. var. *horridum,* Wimm. & Grab. Linnaeus identified the plant; it was moved to the genus *Cirsium* by Johann Scopoli; Wimmer & Grabowski added the variety, which should need no explanation!

Foamflower, Coolwort
Tiarella trifoliata

PLANT PARTICULARS

Many familiar wild plants are not natives. Introduced, non-native, exotic and adventitious refer to plants foreign to the areas where they now grow. Roads built to extend the Roman Empire helped spread plants, as do modern highways and means of travel; foreign plants usurp space earlier occupied by less aggressive natives. Seeds in personal effects of explorers or traders grew far from their points of origin. Plantain probably came to North America by such an accidental introduction. Recognizing this, Indians called the Broad-leaved Plantain White Man's Foot. Later, plants brought by settlers moved to wild surroundings. Foxglove grew in the garden of Margaret Watkins Gibbs, who came from New York to Southern Oregon in 1854 as the bride of Addison Gibbs, later Oregon's second governor. When Foxglove escaped, neighbors called it Gibbs' Weed, showing how a common name can start. Wild Pea, Tansy Ragwort, and Bittersweet Nightshade have become much more common locally in recent years. Garlic Mustard, identified in Tryon Creek Park in 1971, pointed out to me as a curiosity in 1977, is now widespread there and will no doubt continue to travel.

Indians had no common oral language, no written language, few land roads to help in exchange of customs and products. For many Northwest natives, highways were in large part waterways. Because their civilizations were many and varied, knowledge about their customs and uses of materials is limited. Their respect for nature and consistent conservation practices provide inspirational examples.

Principal foods of Northwest Indians were fish, roots, berries, and game. Roots were also used for barter and trade. Gathering plant foods was always the work of women. Generous use of plants helped balance their diet. Stems were usually eaten raw after removing tough outer skin. Leaves were eaten raw or cooked and also provided beverage or medicinal teas. The cambium layer of trees, (Latin *cambium,* a nourishing body fluid), between outer bark and inner wood, was an important food at times, often considered a survival food, sometimes pounded and dried for winter use. Its soft tissue provided a rich supply of nutrients, including minerals and vitamins. Indian cooks often flavored stews or root soups with dried berries. They found meat cooked up faster and was more tender when cooking liquid included one of the Docks. They learned to change cooking water several times to leach out strong flavors or possible poisons in wild plants used for food. Acorns and Skunk Cabbage require such treatment.

Berries were eaten raw or cooked, dried for storage, or pounded with meat and/or fat as pemmican. Coast Indians often ate berries—Huckleberry, Salal, Service, Thimble, Salmon, Gooseberries—with seal oil. Pioneers and settlers ate berries extensively, as we do today, raw or cooked, in pie, jelly, preserves, sauce, and wine.

Ferns provided food, medicine, and materials. The uncurling fiddleheads of some were eaten, raw or cooked, when 4 to 7 inches high. Roots were roasted and peeled; starchy material separated from fibrous parts was eaten, or dried for use as flour. Some groups dug fern roots in the fall; others ate these roots in the spring. Fern fronds were used as we use paper towels, foil, or wax paper: to wrap, cover, or wipe food, to lay out fish or other food for cleaning. Ferns also served to line pits for cooking, and for mattresses or camp mattresses.

More common plants were extensively used; what some groups used for food or medicine might be thought poisonous by others. No doubt native Americans became accustomed to foods which would disagree with us. Starvation food notations confirm that Indians ate some plants because they were available rather than particularly desirable. Some uses fit familiar concepts. Spitting chewed Salal leaves on burns agrees with what we have been taught because these leaves contain tannic acid. (Have you heard of a teabag as a home remedy for burns?) Eye medications were important in a culture where poorly vented wood fires were major methods of cooking and heating. Many shampoos and hair treatments were mentioned. Herbal dyes vary appreciably with the season and parts used; wool is easier to dye than cotton.

Food, materials, and medicinal uses of some Northwest wild plants are listed alphabetically by common names, followed by scientific names with synonyms in parentheses. Blooming periods are approximate for the Portland area.

Alder
Alnus rubra
pinnate leaf veins, notched leaf edges

ALDER
Alnus rubra (A. oregana),   Betulaceae

NAMES   Alder may be from the Anglo-Saxon *alr. Alnus*, the ancient Latin name, and *rubra*, Latin for red, combine in the scientific name; freshly cut sapwood is red. Alder blooms in February, permitting pollenization before leaves interfere. The reddish male flowers, catkins, hang down. Staminate (female), flowers develop in woody strobiles or cones, which often hang on the trees after seeds have been released. Alder likes well-drained moist locations where its roots will not be continually wet. Working with or through bacteria on the roots, Alders are able to fix nitrogen, which occurs naturally in soil air in a form otherwise unavailable for use.

FOOD   Several British Columbia coastal groups, Sechelt, Mainland Comox, and some Coast Salish groups, ate Alder cambium fresh, or dried it in cakes for winter use. Most Washington state Indians apparently did not eat Alder cambium, though Swinomish ate sap scraped from cambium taken only when the tide was coming in. Thompson and Saanich used Alder bark or leaves in steam-cooking pits to color onions or Camas bulbs red.

MATERIALS   Oregon's most important hardwood, Alder is valued for furniture and useful for open fires as it does not spark. Indians traditionally preferred Alder for cooking salmon. Next to Cedar, Alder was most widely used by Northwest coast Indians for woodworking, including dishes, spoons and platters. Quinault made the fire drill, canoe bailers and canoe paddles of Alder, though Swinomish regarded it as too soft for paddles. Quileute seasoned green Alder to make paddles which would not split in the hot sun. Swinomish and Quinault stored Elderberries in containers lined with Alder bark. Sunset Magazine recently recommended Alder for furniture parts, cabinetry and plywood.

MEDICINE   Swinomish treated colds and stomach trouble with an Alder bark decoction. Quileute used raw cones to treat dysentery, while Klallam chewed catkins to relieve diarrhea. To ease aching bones, Cowlitz rubbed rotten Alder wood on the body. North American Indians used a bark decoction to reduce pain from burns or scalds. Herbal medicine has used bark of *Alnus* species as alterative, tonic, astringent, emetic, and bitter, and in an ointment for skin disease. Parkinson in 1640 wrote of using Alder leaves to attract fleas, which were then swept from the house with the leaves.

OTHER   Alder bark provided a brown dye for tanning leather. Quinault, Snohomish, Lummi, and Quileute found this dye helped make nets invisible to fish. Other dyes from Alder vary from bright red to orange and dark brown, depending on treatment used.

Quick growth lessens danger of soil erosion, so this weedy tree is useful in planting wastelands, fills and dumps. Life span of Alder is relatively short, about 50 to 60 years; it grows 70 or more feet tall. Geological records show Alder to be an ancient genus.

SMALL-FLOWERED ALUMROOT
Heuchera micrantha   Saxifragaceae

NAMES   Alumroot reports the astringent taste, like alum. Johann H. Heucher was a German botanist. *Micro,* small, and *anthemon,* flower, describe the delicate white flowers which appear in April and May. The basal leaves have long hairy stems.

MEDICINE   Skagit treated cuts with a root solution, (a use for *H. americana* by Indians of eastern North America), and soaked the whole Alumroot plant in water used to wash the hair of little girls, to make it grow thick. Indians of Nevada and Utah treated heart disease with a root decoction. Herbalists have recommended roots of *Heuchera* species to treat dysentery and diarrhea, and the finely ground root as a first aid treatment for cuts and abrasions. Eating the raw root was an old folk remedy; Kirk suggests campers be familiar with this genus in case of diarrhea problems. A root tea has been used as a gargle for sore throat. An eastern Alumroot, *H. americana,* was included in the U. S. Pharmacopoeia in 1820-22, and again in 1850 as an internal or external astringent.

OTHER   This native perennial relative of Coral Bells, *H. sanguinea,* is recommended for wild gardens.

OREGON ASH
Fraxinus latifolia (F. oregana)   Oleaceae

NAMES   This common name came from the Anglo-Saxon *aesc.* Ash splits easily; the Greek *phraxo,* to split or cleave, appropriate for the genus, is also the source of the ancient Latin name for the Ash tree. *Latifolia* says broad leaved. Ash is the only native Oregon tree with compound leaves; the five or seven leaflets help identify it. It grows rapidly in moist areas on river banks or low-lying land, blooming from March to May; pollen and fruit grow on separate trees. Winged Ash seeds or samaras are single; those of Maples are always in pairs. Many Ash trees are from 100 to 150 years old; specimens 250 years old have been known.

MATERIALS   Cowlitz and Quinault used Ash for canoe paddles, Cowlitz for digging sticks. Some groups favored Ash for war spears because of its weight and strength. The strong wood is used to manufacture sporting equipment, as handles for small tools, and for fence posts. Ash provides shade and fuel, and is remarkably disease-free; the bark often supports lichens. It is browsed by elk and deer.

MEDICINE   Cowlitz drank a bark decoction for worms. Stuhr said Indians treated wounds received in bear hunts with roots of Oregon Ash. Herbal medicine recognized dried Ash bark as diuretic, tonic, alterative, and laxative. The National Formulary formerly included dried bark as diuretic and tonic. Of some forty *Fraxinus* species, only one, *F. latifolia,* grows in our area. Species medicinally used are *F. excelsior,* common European Ash, and *F. americana,* White Ash of

Canada and eastern United States, but some sources also add "and probably other species!" Dried bark was used most, but leaves were sometimes used. Pollen of *Fraxinus* species is a cause of hay fever and contact dermatitis.

OTHER   Oregon Ash has been recommended for revegetating wet, low-lying areas. Bark of the European Ash has been used in tanning. Mountain Ash, a Rose family member, is not related to this tree.

DOUGLAS ASTER
Aster subspicatus   ~~Names~~ *Compositae* RBW.

NAMES   The Greek *aster*, a star, describes the purple, somewhat spiked, *subspicatus*, flower heads. This pretty perennial blooms from July in open woods or woodland edges.

OTHER   Some find touching the flowers causes an allergic reaction or contact dermatitis. This plant may be dangerous to livestock in areas where the soil contains selenium, as its roots absorb the selenium, and eating the plant could cause a problem.

LARGE-LEAVED or YELLOW AVENS
Geum macrophyllum   Rosaceae

NAMES   Avens is from the Old French *avence*.   Large-leaved reports size of the end lobe of the basal leaves, as does the species name, *macrophyllum*.   The stem leaves are smaller and are often divided into three parts. *Geyo* means to taste well in Greek; roots of some species are edible. Yellow flowers of this perennial native

Larged-leaved Avens
Geum macrophyllum
basal leaf

brighten scattered plants in open woodlands from April through October. The petals lack the varnished look of Buttercups, which they somewhat resemble. Hooked seeds take advantage of passersby to spread the species, a technique developed by successful survivors in the plant world.

MEDICINE   Quinault rubbed mashed leaves on open cuts; Quileute and Klallam chewed *Geum* leaves during labor. Chehalis drank a leaf tea as a contraceptive. Herbal medicine used herb and roots of *Geum* species as astringent and tonic, recommended chipped roots boiled in water or milk for dysentery. Tea from herb or root has been used for inflammation of the stomach lining or as a gargle for sore throat. Geum roots are high in tannin.

OTHER   Toward the end of the 13th century many architectural designs used *Geum* plants. Folklore recommended *Geum* roots in the house to make it safe from the devil.

BANEBERRY, CHINA BERRY
Actea rubra (A. arguta)   Ranunculaceae

NAMES   Baneberry refers to toxic qualities, China Berry to the glossy berries. The old Greek name for Elder, *aktea*, reports a belief that the leaves resemble those of an Elder. Some berries are red, *rubra*. Red- and white-berried plants share the same scientific name. The fluffy white flowers stand out above the lacy leaves in May, but the berries, ripe in July and August, are more showy.

MEDICINE   Quileute spit chewed Baneberry leaves on boils to bring them to a head; Quinault used the same treatment for wounds. One herbal recommended roots of a genus member, another, medicine from berries; it was used with extreme care because of possible adverse effects on the heart. Dried ground roots used externally as an acrid irritant caused blistering if too strong. Traditional medicine formerly used the root as an emetopurgative, considered especially dangerous in overdose. These uses confirm that Western Baneberry is toxic, especially berries and roots, though fatalities are rare.

OTHER   The delicate foliage and shiny berries make this perennial attractive for woodland gardens. If small children frequent your garden, remember the berries are poisonous.

BEDSTRAW or CLEAVERS; FRAGRANT BEDSTRAW
Galium aparine; G. triflorum   Rubiaceae

NAMES   Bedstraw has been used to stuff mattresses. One name, Our Lady's Bedstraw, points out use by ladies of rank—or perhaps reported use in the manger of the infant Jesus. Some genus members were used to curdle milk for cheese, hence *gala*, Greek for milk. The Greek *aparo*, to seize, refers to the plant's climbing habit, as does the name Cleavers. Backward-pointing bristles on the stem of the European native *G. aparine* help the plant climb above ground level. Bristles of other species are less prominent but still effective for

34

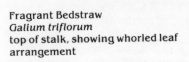

Fragrant Bedstraw
Galium triflorum
top of stalk, showing whorled leaf
arrangement

Bedstraw
Galium aparine
whorled leaves

climbing. Bedstraws have square stems, whorled leaves, and hitch-hiking seeds with stiff hooked hairs. The tiny white flowers bloom from April until June. Number and size of leaves in whorls, and number of flowers in a cluster vary in different species. The *triflorum* of the native Fragrant Bedstraw refers to the usual combination of three flowers on its blooming stems.

FOOD   Coffee is a member of the Madder family. Dried, roasted Bedstraw seeds have been used for a coffee substitute, and the dried plant for a tea. It would take a great many of the small paired seeds to make much coffee. Euell Gibbons suggests *G. aparine* as a potherb when very young, though other wildfood sources do not mention it.

MEDICINE   Northwest Indians used the native Fragrant Bedstraw. Quinault, Makah, and Klallam rubbed mashed plants on their hair to make it grow; Snohomish and Lummi rubbed the body with Bedstraw for a good smell; the plant smells especially sweet after it wilts. A charm used by the Cowlitz provided that a woman who used the proper incantations and rubbed Bedstraw on her body while bathing would be successful in love. Herbal medicine has recommended Bedstraw tea as a diuretic, for urinary tract problems, and for tonic, alterative, and aperient qualities. Culpeper recommended *G. aparine* as a spring tonic. Scalds and burns have been treated with a plant ointment, and slowly healing burns bathed with plant juice or a strong tea.

OTHER   Cowichan rubbed Bedstraw plants on their hands to remove pitch. Dried plants have been used as tinder for fire starting. *Galium* roots produce a red or purple dye, leaves, a purple dye. The native, mostly perennial, Bedstraws are considered better for dyes, the introduced annual, *G. aparine*, more desirable for medicine. White-tailed deer, ducks and geese eat Bedstraw.

Sweet Woodruff, *Asperula odorata*, sometimes escapes to the wild west of the Cascades. Hortus Third calls this *Galium odoratum*.

BISHOP'S CAP, LEAFY MITREWORT
Mitella caulescens   Saxifragaceae

NAMES   Seed cap shape gives genus and common names; *mitella*, the diminutive of the Latin *mitre* or cap. *Caulis*, Latin for stem, says the flower has a stem. Greenish-white flowers of this native perennial bloom from the top of the stem down, appearing from late April in deep woods to moist meadows.

OTHER   Bishop's Cap is a good groundcover in shaded gardens. Its delicate stamens, reduced to mere threads, will challenge photographers.

LITTLE WESTERN BITTER CRESS
Cardamine oligosperma   Cruciferae

NAMES   The Greek *kardia*, heart, and *damao*, to strengthen, provide the genus; *oligosperma* means few seeded. This annual or biennial native weed of woods and gardens blooms uninvited in February and March.

FOOD   Kirk lists 15 *Cardamine* species as edible in a less than enthusiastic recommendation; many non-poisonous plants do not taste particularly good. Wood Bitter Cress, *C. angulata*, a perennial native, blooms in April.

BITTERSWEET NIGHTSHADE, SMALL BITTERSWEET
Solanum dulcamara   Solanaceae

NAMES   When chewed, root and stem taste first bitter, then sweet, hence Bittersweet. The species name reverses this, *dulca* being bitter, *amara*, sweet. The Latin *Solanum*, I ease, recognizes the plant's mildly narcotic tendency. This Eurasian native, within my memory a curiosity here, is now widespread, as birds have enjoyed the fruit of ornamental plantings. The handsome flowers with turned-back purple petals and yellow centers appear from May to September. Flower clusters and ripe red berries may both be present as the season advances.

FOOD   Some sources list berries and leaves as poisonous; Hitchcock says herbage and fruits are mildly poisonous to livestock and humans. Ripening of the fruit is believed to lessen the poison, hence unripe berries are more dangerous. Berries have been used for pies and sauce; cooking apparently destroys the alkaloid solanine in the berries. As with some medications, reaction to eating (raw) fruit seems to be proportional to body weight of the consumer. This could explain why small children have problems with these pretty berries. How many little ones eat any attractive—especially red—fruits they can find?   This is another reason to warn: do not eat anything not positively identified as non-poisonous and edible.

MEDICINE   Traditional and herbal medicine formerly used Bittersweet Nightshade. Branches from plants one or two years old, collected after the leaves had fallen, were used to treat chronic rheumatism, bronchitis, and various skin diseases. Before 1907, *S. dulcamara* was in the British Pharmacopoeia; dried stems were formerly in the U. S. National Formulary. More recent references say its use has been almost completely discarded.

OTHER   The Potato and Tomato have flowers similar in shape and are from the same plant family. Deadly Nightshade, *Atropa belladonna*, also in the Nightshade family, is the source of the narcotic drug atropine, used for drops to dilate eyes. The dirty-purplish to greenish or yellowish bell flowers of this European native are usually solitary; its black berries contain alkaloids which can be fatal. It is occasionally found as a weed west of the Cascades, but does not commonly grow wild in our area.

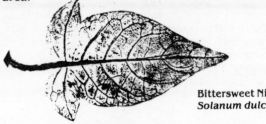

Bittersweet Nightshade
Solanum dulcamara

DEWBERRY, WILD BLACKBERRY

Rubus ursinus (R.macropetalus, R. vitifolius)   Rosaceae

NAMES   The Latin *rubus*, bramble, combines with *ursus*, bear, for this currently-accepted name. My favorite native berry is the fruit of this creeping vine, which blooms from April through July. Male and female flowers grow on separate plants as this is a dioecious perennial.

FOOD   All Northwest Indians used the Dewberry; some ate it fresh, others also dried mashed berries in cakes for winter. Meriwether Lewis reported that Indians along the Sandy and Columbia rivers ate only fresh berries. These make lovely pies, jams, and jellies. Although technically the plant is deciduous, losing its leaves in winter, the old reddish leaves often hang on. Saanich, Sechelt and Comox boiled the red leaves for a refreshing beverage. Tea can be made from fresh or dried leaves. Cowlitz and Quileute used both leaves and vine for tea.

MEDICINE   Skagit treated stomach trouble with a leaf tea. Herbalists recommend Blackberries for treating diarrhea, using roots, leaves, bark and fruit as medicinal parts. They are rich in tannin, astringent, and tonic. Roots of some eastern Blackberries were formerly included in the U. S. Pharmacopoeia and the National Formulary. Medicinal value of dried bark and rhizome is chiefly dependent upon tannin content.

OTHER   The abundantly-armed native Wild Blackberry takes root at intervals, providing good erosion control on burned and cutover lands. The same persistent growth makes it undesirable for gardens. Use in hybridizing such fruits as the Loganberry, Youngberry and Boysenberry was no doubt due to the excellent flavor of these berries.

EVERGREEN BLACKBERRY; HIMALAYA BLACKBERRY

Rubus laciniatus; R. discolor   Rosaceae

NAMES   *Laciniatus*, Latin for torn or fringed, describes the finely cut dark green leaves of the perennial Evergreen Blackberry. Three synonyms for the Himalaya Blackberry, *R. procerus, R. thyrsanthus,* and *R. fruticosus*, indicate problems of plant classification. Birds which enjoyed these European natives in gardens distributed them. They enjoy their own company, often growing in large patches from which they continue to spread, most commonly west of the Cascades. Evergreen and Himalaya Blackberries are widely used. These ripen in August and September and are easier to pick in quantity than their smaller native cousins, also far less flavorful.

OTHER   Berries with alum as a mordant make a bluish-grey dye, and young sprouts, mordanted with iron, make a black dye. The Evergreen Blackberry is somewhat less hardy than the Himalaya, though both are widespread weeds, difficult to control.

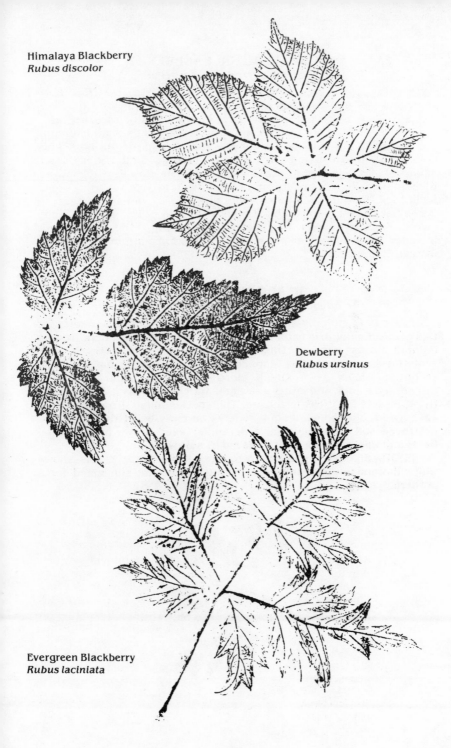

Himalaya Blackberry
Rubus discolor

Dewberry
Rubus ursinus

Evergreen Blackberry
Rubus laciniata

BLACKCAP, BLACK RASPBERRY
Rubus leucodermis   Rosaceae

NAMES Blackcap and Black Raspberry describe the fruit. Both stem and leaf undersurface of this shrub which blooms in April and May are silvery white, noted by the Greek for white skin, *leuco, dermis*.

FOOD Klallam ate sprouts and young leaves in the spring. Comox ate peeled young sprouts. Calapooya and Molalla ate spring shoots raw, ate berries fresh and dried them for storage. Straits Salish, Stali, Squamish, Sechelt, Comox, Kwakiutl, Bella Coola, Cowlitz, Green River, Klallam, Puyallup, Nisqually, and Green River ate fresh berries; Upper Stali dried them to use like raisins. Cowlitz dried and stored them in Maple bark baskets for winter. Green River also dried them. Puyallup mixed Blackcaps with Blackberries for drying; they sometimes included these when cooking meat. Today berries are made into jam instead of dried cakes.

BLADDER CAMPION
Silene cucubalis   Caryophyllaceae

NAMES Campion, Catchfly and Wild Pink are common names for this genus; *Campion* is Latin for campus or field. Bladder identifies the inflated flower base of this species. Silenos was a Greek god, sometimes represented as foam-covered; some species have a sticky secretion; *sialon* is a Greek word for saliva. Plant stickiness discourages ants and other crawlers, which are less efficient than flying insects for cross fertilizing. The species name, from the Greek *kakos*, bad, plus *bolos*, a shot, reflects an early belief that the plant destroyed soil. This aggressive perennial weed grows in cultivated fields and waste places; deep roots aid its survival.

FOOD Bladder Campion, introduced from Eurasia, has showy white flowers in June and July. Young shoots are suggested for a potherb.

Bleeding Heart
Dicentra formosa
top of leaf stalk

WESTERN or PACIFIC BLEEDING HEART
Dicentra formosa   Fumariaceae

NAMES Heart-shaped pink flowers of this lacy perennial gave the common name; look for them in moist woods in March and April. *Dicentra* comes from the Greek *dis*, twice, plus *kentron*, a spur, as the corolla of this flower has two spurs; *formosa* is Latin for beautiful.

MEDICINE Upper Skagit boiled the pounded root of this native for worm medicine. Skagit chewed roots for toothache relief, used the crushed plant on the hair, and soaked the plant in water for a shampoo to make hair of young children grow.

OTHER The fruit, shaped somewhat like a garden pea, is toxic. All parts contain alkaloids poisonous to livestock which eat it. A small oil appendage on the seed attracts ants which spread the plant as they scatter seeds collected for food. Bleeding Heart, a favorite in woodland gardens, causes contact dermatitis for some people.

BOG or REIN ORCHID
Habenaria dilatata   Orchidaceae

NAMES Bog is the preferred habitat; *Habenaria*, from Latin *habenas*, strap or rein, giving genus and common names, describes the lip shape. *Dilatata*, expanded, refers to the spread-out base of the Orchid lip.

FOOD The tuber-like roots are edible, raw or cooked, but should be considered for emergency food only. Enjoy this fragrant native North American white Orchid in boggy places where it blooms in June and July, remembering that its needs are so specific that transplanting is rarely successful.

BRAKE or BRACKEN FERN
Pteridium aquilinum   Polypodiaceae

NAMES Brake or Bracken may have come from the broken appearance of ferns after the first frost. *Pteridium* is from the Greek *pteris*, fern, or *pteron*, wing, perhaps because leaf divisions of Brake Fern are somewhat winglike. It is difficult to explain *aquilinum*, Latin, *aquila*, eagle. Brake, most widely known and most common of our ferns, most valuable by far of edible ferns used by Northwest Indians, grows world-wide in the northern hemisphere.

The single stalks with three-forked, broadly-triangular leaves spring up here and there in open woodlands and fields from underground rootstocks. Brake Fern spores grow under the edges of the fronds, though not all have them. Roots may stretch twenty feet from point of origin and be five inches to two feet or more below the surface. Bracken frustrated settlers, as roots often went deeper than their plows, and ferns continued to grow, fed by starch stored in the undisturbed root systems. Cut roots sprout to start new plants.

Brake Fern
Pteridium aquilinum
tip of frond

FOOD Brake Fern has been eaten in many parts of the world, uncurling fronds or fiddleheads raw or cooked, roots cooked, often made into bread, especially in times of famine. Young sprouts, picked when six or eight inches tall, are eaten as salad or cooked. In Japan this fern received legal protection needed because of its popularity as a food. In addition to eating fiddleheads, Japanese thickened and flavored soups with the root. Modern sources consider this an emergency food, not flavorful enough to recommend for general use. Sturtevant reported that the rhizomes were combined with malt for brewing beer in Siberia. Recent studies suggest that Brake Ferns, especially raw, have carcinogenic qualities.

Northwest Indian groups ate Brake Fern extensively. Cowlitz ate young sprouts raw; some British Columbia groups ate boiled fiddleheads. Almost all coastal British Columbia groups ate rhizomes. Lummi, Skokomish, and British Columbia groups dug roots in fall or winter. Lewis and Clark spoke of wide use of Brake Fern for food by Indians whom they visited, including bread made from root flour. Roots were baked over hot stones in a pit; water was sometimes added. Indians pounded roasted roots to remove fibers and ate the starchy remainder as a paste or as flour. Pounding sticks, important tools for Indian women, were often made of Yew. Snohomish stored baked rhizomes in baskets.

MEDICINE Snohomish ate raw rhizomes for bronchial trouble. Some California Indians used Brake Fern rootstocks for diuretic and worm medicines.

OTHER Quileute burned over prairies to encourage growth of young shoots to lure elk and deer. Indians used Brake Fern for wiping and covering foods, for lining cooking pits, and for mattresses and camp mattresses. Brake Fern provides valuable shelter for small game. Dried fronds included in hay are poisonous to stock. The poison is cumulative; it causes a vitamin B deficiency in horses and affects the bone marrow of cattle.

BUCKBRUSH or OREGON TEA
Ceanothus sanguineus   Rhamnaceae
NAMES Common names for this May- and June-blooming shrub report use as deer browse and as tea. *Ceanothus* is a Latin version of the Greek *keanothos,* ancient name of some kind of thistle; some species have spiny branches. *Sanguineus,* Latin for blood red, recognizes the reddish color of young branches, as does the name Redstem Ceanothus. Buckbrush leaves have three prominent veins starting from the leaf base, a characteristic of the *Ceanothus* genus.

FOOD Flowers and leaves of all species can be used for tea; some taste better than others. One provided substitute tea following the Tea Party of the Revolutionary War. Sweet said California Indians ate the seeds.

MEDICINE Kootenay used Buckbrush tea as medicine for tuberculosis and as a beverage. Traditional and herbal medicine formerly used bark and roots of Ceanothus species as astringent and tonic, for coughs and sore throat; pounded leaves were used as a poultice for burns or to dry up sores. A lather of crushed fresh flowers in water is said to leave the skin soft.

OTHER *C. sanguineus* is well worth growing in home gardens. Leaves have been used as a tobacco substitute, and a red dye has been made from the roots. Okanagan used this shrub to smoke deer meat when other wood was not available. Some blue-flowered *Ceanothus* are called Wild Lilac.

BUGBANE
Cimicifuga elata   Ranunculaceae

NAMES Both common and genus names suggest former insect-repellent use of some genus member, from *cimex*, bug, and *fugare*, to drive away. Latin *elata*, tall or elevated, describes this native perennial which blooms in June and July in moist woods west of the Cascades.

MEDICINE Herbal medicine has used an eastern species. *C. racemosa*, as alterative, emmenagogue, diuretic, and stimulant. *C. racemosa* was an ingredient in the popular American patent medicine, Lydia Pinkham's Vegetable Compound. Traditional medicine included root and rhizome in the U. S. Pharmacopoeia from 1820 to 1936, and the National Formulary from 1936 to 1950, as sedative, emmenagogue, and as a treatment for rheumatism. A medical dictionary of 1956 says there is no evidence of their therapeutic value.

OTHER Bugbane has been recommended for a wild garden. It is related to the Baneberry.

COMMON or LESSER BURDOCK
Arctium minus   Compositae

NAMES The "bur" in the common name tells of the many hooks on the seeds; Burdocks are related to Thistles. The genus name comes from the Greek for bear, *arktos. Minus* means smaller; this species is smaller than the somewhat similar *A. lappa*, Great Burdock, occasionally found west of the Cascades.

FOOD Burdock has been cultivated in Europe and Japan as food. Wildfood sources recommend year-old roots of this Eurasian biennial for salad and potherb; older roots become woody. Changing cooking water on roots and leaves lessens the strong flavor. Ground, roasted roots have been used as substitutes for tea or coffee.

MEDICINE Herbal and traditional medicine have used roots and seeds of Burdock as tonic, diaphoretic, alterative, diuretic, and blood purifier. Tea from seeds has been used for skin conditions, and a root solution as a wash for burns, wounds, and skin irritations, helpful because it contains tannic acid. The dried root of *A. lappa* was included in the U. S. Pharmacopoeia from 1831 to 1842 and from 1851 to 1916 and the National Formulary from 1916 to 1947. The Dispensatory of 1950 said it is no longer considered to have medicinal value.

Woods Buttercup (r.)
Ranunculus uncinatus

Creeping Buttercup (l.)
Ranunculus repens

CREEPING BUTTERCUP
Ranunculus repens   Ranunculaceae

NAMES Creeping, Latin *repens*, is accurate, for plant stalks send down roots where they touch the ground. Buttercup reflects the old game of holding the shiny yellow flower under a child's chin to see whether the child likes butter. *Ranunculus* means little frog; frogs like the moist areas where many Buttercup plants grow. This species, with shiny petals and abundant pollen, blooms from May to August or later. Petals of Buttercup flowers typically have a glossy surface, as though varnished.

FOOD Gunther reported Makah and Quileute ate roots of some Buttercups, dug in fall and winter, and cooked on hot rocks. These were dipped in whale or seal oil and eaten with dried salmon eggs. Some California Indians ate Buttercup seeds, parched and beaten to use as flour. Current wildfood books recommend against food use of any Buttercup.

MEDICINE Former use of *R. repens* as a counterirritant was infrequent because of the violent reaction. European beggars used to solicit pity by displaying ugly sores they had produced with the irritant Buttercups.

OTHER This introduced European perennial, more common west of the Cascades, crowds out grasses and other native plants, helped in part by refusal of stock to eat the bitter-tasting Buttercup when other food is available. Many Buttercups are poisonous to livestock, especially cattle, toxicity varying with species and stage of growth. The glycoside ranunculin contains an unstable oily irritant poison, the volatile part of which is driven off by drying, as hay is safe for stock. This, as many *Ranunculus* species, causes dermatitis for some people.

Buttercup flowers provided yellow dye to Indians. Leaves of *R. repens* are rarely a solid green but are usually somewhat blotchy; one source said they were faintly marked with white. It is a challenge to find plants with color or shade variations on one leaf surface, or different colors on leaf top and under side.

WOODS BUTTERCUP
Ranunculus uncinatus (R. bongardii)   Ranunculaceae

NAMES Growth pattern shows this is a woodland plant, for the rangy herb stretching toward the light could not withstand the heavy winds of open areas, where plants are often protected by low, compact growth. Another name, Bongard's Buttercup, honors a Northwest botanist. The Latin *uncinatus*, hooked at the point, fits the Woods Buttercup; the very pronounced hooks of the dry one-seeded fruits arrange free transportation from passing animals and people and spread the species. The inconspicuous yellow flowers are so small that it takes a close look to recognize the shiny petals typical of Buttercups. This annual or perennial native is neither invasive nor very attractive.

MEDICINE Makah mashed leaves of *R. uncinatus* and covered them with a small shell as a poultice to prevent blood poisoning.

CALIFORNIA POPPY
Eschscholzia californica   Papaveraceae

NAMES   Poppy comes from the Celtic *papa* meaning pap, because Poppy juice was formerly put in children's food to make them sleep. *Eschscholzia* honors Johann Friedrich Eschscholtz, a physician and naturalist who visited California and the Northwest coast with Russian expeditions of 1816 and 1824. These cheery orange flowers respond to light, opening early in the day and closing early or in cloudy weather. Where abundant, California Poppy provides good winter forage for cattle. More common farther south, these perennials were little used in the Northwest.

MEDICINE   Some California Indians used *E. californica* leaves for toothache relief and other medicine.   Spanish-Californians formerly fried California Poppy plants in olive oil and added perfume to make a hair oil thought to promote growth of hair and make it glossy. The U. S. Dispensatory, 1950 edition, says *E. californica* does not have the pain-relieving virtues attributed to some Poppies, as its alkaloids have only feeble effects.

CANADA DOGWOOD, BUNCHBERRY
Cornus canadensis   Cornaceae

NAMES   Bunchberry describes the berry cluster which follows the flowers on the green leaf-platform of this low subshrub. It grows over much of Canada, *canadensis*.

FOOD   The orange-red berries are reported edible but insipid in flavor.   Some wildfood authors suggest straining out the large fruit stones and combining the pulp with other berries for jelly or conserve. Washington Makah and several British Columbia coastal groups ate berries raw, with grease.   Haida sometimes steamed and preserved them for winter in water and grease.

OTHER   This is a good groundcover or ornamental in woodland gardens, though snails enjoy it, too. It grows well in soil which includes decaying wood. The actual flowers, from May, are tiny; white modified leaves provide the showy effect.

CANDY FLOWER, SIBERIAN MINER'S LETTUCE
Montia sibirica   Portulacaceae

NAMES *Montia* honors an Italian botanist.   For other name comments, see p. 20

FOOD   Wildfood lovers nibble Candy Flower leaves, stems, flowers and roots raw, confirming the Lettuce of the common name, and cook these as spinach. It provided miners a welcome vegetable, a good source of vitamin C, which they needed to prevent scurvy. This prolific annual herb is an easily-recognized and readily-available edible plant, which grows and blooms over an extended season. It is more common than Miner's Lettuce, *M. perfoliatum*.

MEDICINE Snohomish, Quileute, Skykomish and Cowlitz rubbed Candy Flower stems between the palms, rubbed the plant in cold water, and used this as a tonic to make hair glossy. Quileute used the same product as a dandruff preventive. Quileute used a plant tea as a urinative and rubbed juice of the stem on the eyes. Skagit drank a plant tea as a general tonic or to treat sore throats. Quinault women chewed the whole plant during pregnancy to assure that the baby would be soft when born.

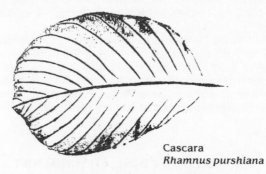

Cascara
Rhamnus purshiana

CASCARA, CHITTIM, BUCKTHORN
Rhamnus purshiana   Rhamnaceae

NAMES Spanish-Californian priests called this holy bark, *Cascara Sagrada*, because local Indians held it sacred. *Rhamnos*, Greek name for Buckthorn, may have come from the Celtic *ram*, a branch or tuft of branches. Flowers, from April, are a good nectar source, providing another common name, Bee Plant. The blue-black berries are a favorite food of robins, grouse, wild pigeons, raccoons and other wildlife. The hard tough seeds pass through their digestive systems undamaged.

FOOD and MEDICINE Nootka and Makah ate berries fresh in July and August. Universally used as a laxative, Cascara bark has been commercially important in the Northwest because Oregon, Washington, and Northern California were major supply sources. Cascara is one of the most important natural drugs produced in North America, and one of few plant medicines still used. It has long been listed in both United States and British Pharmacopoeias. A European and North African species, *R. catharticus*, has similar laxative properties, as does *R. californica*. Of about 100 Rhamnus species, only two grow in Oregon; *R. alnifolia*, Buckthorn, grows on the eastern side of the Cascades. Bark for commercial use is aged from one to three years to avoid the emetic action of green bark. Its laxative properties were important to native Americans whose diets included such constipating foods as Acorns and Brake Fern. It was known and used by all. British Columbia groups valued Cascara as a tonic and laxative, but it was less available in their territory.

Squaxin washed sores with a bark infusion, or spit chewed bark on sores. Skagit burned bark, mixed the charcoal with grease, and rubbed this mixture on swellings. Herbalists have recognized Cascara in small doses as stomachic, tonic, and bitter, promoting digestion and appetite. Cascara has been much used in veterinary practice, especially for dogs.

OTHER   It is important that those who need hotdog roasting sticks be able to recognize and avoid using Cascara, which has a leaf somewhat like that of Alder. Cascara leaves are darker green than those of Alder, a paler green below, with finely-toothed leaf margins, are somewhat more oblong in shape, and have more prominent veins on the leaf back. Cascara causes contact dermatitis to some people. A Cascara specimen was collected in Idaho by William Lewis of the Lewis and Clark expedition; *R. purshiana* grows chiefly west of the Cascades.

Skagit used boiled bark for a green dye on mountain goat wool. Cascara bark has been used on cotton to provide a tan or gray, on wool for a light brown or tan. Berries of *Rhamnus* species have provided brown and yellow dyes.

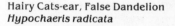

Hairy Cats-ear, False Dandelion
Hypochaeris radicata

HAIRY CATS-EAR, FALSE DANDELION
Hypochaeris radicata   Compositae

NAMES   Cats-ear, a European perennial with hairy leaves, is one of the worst lawn pests; it blooms from May on into October; its wiry flowering stalks resist mowing and often spring back up after the lawn mower has passed. Theophrastus used this name for some Composite. *Hypo* is Greek for under, *chaeris* means young pigs, assuming that pigs like the roots, *radicata*. Cats-ear stems are sparsely branched and do not have milky juice, unlike the Dandelion's unbranched stems with milky juice. Its seeds are well suited for wind dispersal.

A similar annual European weed, *H. glabra* (smooth), has less deeply-cut, less-hairy leaves. Smooth Cats-ear blooms mostly in May and June. Both Hairy and Smooth Cats-ears are more common west of the Cascades. Also similar is the Hawkbit or Fall Dandelion, which has unbranched flower stems.

FOOD   Hairy Cats-ear or False Dandelion, much more abundant than Dandelion, is said to be edible though hairy. An 1862 book about useful British plants reported cultivation of this Cats-ear as a potherb had fallen into disuse.

Hairy Cats-ear, False Dandelion
Hypochaeris radicata

CAT-TAIL
Typha latifolia   Typhaceae

NAMES   Cat-tail describes the fluffy seed heads of this perennial. *Typha* was the ancient Greek name for the plant; *latifolia* means broad leaved.

FOOD   Cat-tail is perhaps the best-known wild edible plant; shoots, roots, and pollen are all recommended.   Wildfood editors include many recipes.   Northwest Indians ate roots raw or baked. Cat-tail is eaten by wild animals, but plants are toxic to livestock if consumed in quantity.

MATERIALS and MEDICINE   Cat-tail made a more important contribution to the Indian economy as a material than as food.   From these Northwest Indians made mats, bags, hangings, screens inside winter houses, coverings for summer shelter and other woven materials. Settlers and Indians used the soft down to stuff pillows and dress wounds and burns. Indians padded cradleboards and baby beds and made string for sewing mats.   Settlers used leaves for rush-bottomed furniture.   Nineteenth-century traditional medicine used Cat-tail root as astringent, emollient, and detergent, the down, for dressing wounds.

OTHER   Cat-tail plants provide an important refuge for waterfowl and small animals, including shelter for nesting marsh birds.   The pollen may be a cause of hay fever.

WESTERN RED CEDAR
Thuja plicata   Cupressaceae

NAMES   Cedar was the Latin name for some tree with fragrant wood; this has red wood. *Thuja* is the ancient Greek name for a similar tree. The unusual adaptation of the evergreen leaves, *plicata*, plaited or folded, gives Cedar branches a flat appearance. An imaginative person can see a series of small butterflies in the backs of the scale-like leaves.

FOOD   Sandy River Indians ate Cedar cambium fresh and sometimes dried it for future use.

MATERIALS   Cedar is a good lumber source, highly regarded for house siding and shingles, a major source of telephone and power poles and fence posts. Remarkably resistant to decay, Cedar was used

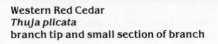

Western Red Cedar
Thuja plicata
branch tip and small section of branch

extensively throughout the Northwest by native Indians. In addition to house planks, the soft but durable Cedar, which splits well and could be used easily before iron became available, provided house posts, roof boards, canoes, cradles, totem poles, boxes, materials for weaving and basketry and for ceremonial uses. Quileute used it for the hearth of the fire drill, as the spindle for spinning mountain goat wool.

For shredding bark, Chehalis women used a deer-bone chopper. Finely shredded Cedar bark was used to pad infants' cradles, as sanitary pads, as towels. Coarser strands were woven into skirts and capes. Groups close to salt water used more Cedar bark clothing; others might use skins. Shredded Cedar wads used as tinder were carried in slow torches on journeys. Unshredded bark was cut into strips for drying berries; bark was used to line cooking pits; several Puget Sound groups laid out boiled salmon on smaller bark mats. Quileute and Makah made Cedar bark sails. Bark mats were made more extensively by British Columbia Indians than by those of Washington state, who liked to get these by trade from their northern neighbors. British Columbia groups also used Cedar bark baskets, boxes and bark itself as trading items. Canoe bailers were made of folded pieces of bark lashed with Wild Cherry bark.

Cedar limbs soaked in water were twisted into rope. Heavier grades of these ropes were used by whale hunting groups—Quinault, Quileute, Makah—for towing home dead whales. Single pliable limbs were used to tie or sew corners of wooden boxes or to tie crosspieces in canoes. Quinault also used Cedar roots for this purpose. Quinault and Squaxin used Cedar limbs for openwork baskets. Finely-split roots of Cedar trees were widely used in Washington for foundation and weaving material for coiled baskets.

MEDICINE Lummi chewed and swallowed Cedar buds for sore lungs; Cowlitz chewed them for toothache, and Skokomish boiled them for a gargle. Skagit and Cowlitz boiled ends of leaves or boiled tips with herbs for a cold or cough medicine. Quinault steeped the ends of limbs and seeds for an infusion to break a fever. Cedar leaves and limbs were used to scrub the body when bathing. Lummi and Skagit used Cedar for purification rituals after contact with the dead or their homes. Chewed Cedar cones have been put in tooth cavities to relieve pain. Sawdust inhalation by workers in paper mills and wood products has caused the allergic response of bronchial asthma.

OTHER Western Red Cedar grows slowly and may live for 800 to 1000 years. It helps stabilize soil near streams, as it prevents silting. It is a good garden ornamental; holiday decorations often include its fragrant boughs. Birds use *T. plicata* for nesting sites.

Western Red Cedar, *Thuja plicata*, Cypress family, is not a true Cedar. True Cedars, *Cedrus* species, belong to the Pine family. One of the best known is the Cedar of Lebanon; others grow in the Himalayas and North Africa, but none are native to the Pacific Northwest.

BITTER CHERRY, WILD CHERRY
Prunus emarginata   Rosaceae

NAMES   Oregon Indians ate the small fruit occasionally, but the bitter of its common name is apparently appropriate; it is also hard to harvest. *Prunus* is the Latin name of the plum; *emarginata*, having a shallow notch, describes the tip of Wild Cherry leaves.   Bees enjoy nectar of this shrubby native tree which blooms from April.   Slit-like pores, lenticels, on Cherry bark help in identification.

FOOD   Some Indians drank a tea made from stems and bark. Numerous birds and small animals like the fruit.   Twigs and buds provide winter browse for deer and elk, though livestock have been poisoned by eating leaves of some Wild Cherries.

MATERIALS   Saanich considered Cherry wood excellent fuel; they sometimes used it for the hearth of the fire drill.   Both Puget Sound and British Columbia groups used Cherry bark, natural color or dyed black, for basket design.   They also used Wild Cherry to tie prongs of fish spears, wrap fish spears, fire drills, and other implements.

MEDICINE   Squaxin mixed Wild Cherry with seeds of Ocean Spray for a blood purifier.   Quinault drank liquid from boiled bark as a laxative, while Upper Skagit and Skokomish used the same liquid to treat a cold.   Lummi chewed bark to aid childbirth.   Leaves, twigs, and fruit seeds contain poisonous glycosides; the cracked pits are especially dangerous.   Bark of an eastern Wild Cherry was formerly included in the U. S. Pharmacopoeia; it was used most frequently to flavor cough syrups.

CHICKWEED
Stellaria media   Caryophyllaceae

NAMES   Chickweed may refer to the fact that wild and caged birds eat leaves, seeds, and young tops of this garden weed. *Stella*, Latin for star, describes the tiny white star-shaped flowers. *Media* means intermediate, a confusing reference.   This spring-blooming Eurasian annual thrives in disturbed soil.

FOOD   Young leaves have been considered a good salad or potherb in Europe from medieval times.   Wildfood books recommend the same uses, noting that Chickweed is rich in iron.   This species is dried for a beverage tea.   Pigs and rabbits like Chickweed, cows and horses will eat it. *S. media* is said to be the most widely-distributed weed in the world; in our area it is more common west of the Cascades.

MEDICINE   Grieve says Chickweed water is an old wives' remedy for obesity, but fails to include instructions! Herbal medicine has used the fresh or dried herb as demulcent and refrigerant, mostly in the form of ointment.

CHICORY, BLUE SAILORS
Cichorium intybus   Compositae

NAMES The common name Chicory and the genus *Cichorium* probably came from the Egyptian *chikourey; intybus* is an early Latin modification of another far Eastern plant name, *Hendibeh,* used also for Endive, only other species in the genus. On summer mornings bright blue Chicory blooms, a logical source of Blue Sailors, brighten roadsides and fields west of the Cascades.

FOOD Virgil, Ovid, and Pliny mentioned eating Chicory, cooked or raw. Herbalist Dodoens, 1616, reported cultivated Chicory leaves were eaten raw or cooked, the roots boiled or roasted. Chicory greens are high in Vitamin A, an excellent source of potassium, and contain Vitamin C, phosphorus and calcium. Roasted malt from *C. intybus,* used as flavoring in breads and cakes, intensifies the flavor of sugar from 30 to 300 times; the product is FDA approved.

For a coffee adulterant or substitute, early Europeans roasted and ground roots dug before flowering time. It is considered better than Dandelion "coffee". An unsuccessful attempt was made to establish Chicory roots as an American farm crop as a coffee substitute; the plant escaped and continues in the wild. Fresh Chicory provides good fodder for horses, cows, sheep and rabbits, though milk of cattle which have eaten it develops an undesirable flavor. It is also a good honey plant.

MEDICINE Herbal medicine has recommended a poultice of bruised leaves for swellings and inflammations, and a Chicory infusion to treat skin eruptions. The long taproot of this perennial, dried, has tonic, laxative, and diuretic action. Chicory roots have been used to adulterate medicinal supplies of Dandelion roots.

WESTERN CLEMATIS, VIRGIN'S BOWER
Clematis ligustifolia   Ranunculaceae

NAMES *Clematis,* from the Greek *klema,* a tendril, is appropriate for this woody twining vine. *Ligustifolia* records Nuttall's opinion that the leaves resemble those of some plant in the genus *Ligusticum,* Umbelliferae family.   This native perennial, having no backbone, borrows support from its neighbors to climb toward light and sun. Its white blooms, male on one plant, female on another, appear from July. The fluffy seeds hang on through the winter, providing easy identification, also nesting material for birds. Western Clematis grows on both sides of the Cascades in Oregon.

MATERIALS Northwest Indians used the tough stems for basketry, for bowstrings, snares, and carrying nets for water bottles, and the strong bark fiber for bags, mats, ropes, capes, and other garments.   Dried vines are currently used for making floral decorations, wreathes, and small baskets.

Western Clematis
Clematis ligustifolia

MEDICINE Indians and settlers chewed bark for sore throat, steeped bark to treat fevers. Poultices of mashed and moistened seeds were used to treat burns. Paiutes of Nevada used the powdered root for shampoo; Warm Springs, Okanagan, and Flathead of Montana used lather made from bark and leaves for general soap and shampoo. Murphey reported Indians used a leaf infusion to treat cuts on horses. Herbal medicine has recommended tea from dried chopped Clematis plants to treat headaches, and a tincture of the fresh plant for a counterirritant.

OTHER Leaves are said to be poisonous if eaten in large quantities; juice is violently purgative. Some *Clematis* species cause dermatitis to a few susceptible individuals.

CLOVER BROOMRAPE
Orobanche minor   Orobanchaceae

NAMES The Greek *orobus*, Vetch, and *anchein*, to choke, provide the genus name for this Mediterranean native which blooms in June in a southeast Portland day camp meadow. Clover Broomrape steals its food, having no green food-producing capacity of its own. These herbs are root parasites, often on legumes; some species grow on Scotch Broom.

FOOD and MEDICINE Kirk says *Orobanche* species plants are edible raw, though better roasted in ashes of a campfire. Herbal medicine has used this genus; the root is considered most active, but the whole plant can be used. It is strongly astringent, and was recommended for poultices and for internal use as laxative and sedative.

CLOVER
Trifolium species   Leguminaceae

NAMES   Clover is a variant of the Anglo-Saxon *claefre*. *Trifolium* reflects the three-part leaves typical of Clover. Red Clover, *T. pratense,* and White Clover, *T. repens,* European perennials escaped from cultivation, are established throughout western North America. The bushy Red Clover, *T. pratense,* has a taproot; *pratense* is Latin for meadow. *Repens* means creeping; White Clover creeps by stolons. Clovers are used in lawn seed mixtures, as a hay crop, and as green manure. Hummingbirds especially like Red Clover; bees prefer White Clover, a good source of fine honey. Both, like other legumes, have root-associated bacteria which enrich the soil, as they fix atmospheric nitrogen.

FOOD   Clover is nutritious and high in protein, but hard to digest raw, as it causes bloating. Cooked roots and leaves are edible; soaking in strong salt water for several hours or overnight improves digestibility. Sturtevant reports that Clover flowers and pods were ground into a powder for food in times of famine in Ireland and Scotland. Mohney says all Clover species provide a valuable survival food. Some California Indians dried and ate roots; some smoked the dried leaves. British Columbia coastal Indians ate cooked leaves and roots of the native perennial Marsh Clover, *T. wormskjoldii;* Makah ate these roots steamed. Pioneers and some Indian groups drank a beverage or tonic tea made from dried Clover flowers; health food stores often have such a tea. Small rodents eat the blossoms and leaves.

MEDICINE   Herbal and traditional medicine formerly used Red Clover, *T. pratense;* flowers were considered sedative, alterative, antispasmodic, and expectorant; an ointment was made to treat ulcers. Although the National Formulary previously included the dried inflorescence, the 1950 Dispensatory stated Clovers are no longer believed to have medicinal virtue.

WESTERN COLTSFOOT
Petasites frigidus (P. palmatus, P. speciosa)   Compositae

NAMES   The name Coltsfoot was borrowed from a European plant which it may or may not resemble. Butterbur refers to early use of the large leaves to wrap butter for storage. The genus of this perennial, from the Greek *petasos*, meaning broad-brimmed hat, refers to shape of the basal leaves. It grows in damp shaded areas, especially on disturbed ground. From early March the purplish-white flower heads of native Coltsfoot, appearing before the conspicuous large leaves, often stand alone on roadside cuts.

FOOD   Mohney says Indians historically burned rolled, dried leaves on a flat stone and used the ashes for a salt substitute, a use still recommended, as the leaves contain sodium chloride. Muckleshoot boiled and ate plant stems. Quinault used the big leaves to cover berries cooked in a pit. Japanese call Western Coltsfoot *fuki*, and eat the stalks; these are parboiled, then sliced and cooked more. Young foliage and flowers also have been eaten as a potherb.

MEDICINE Quileute prepared a tea from the boiled root or ate the raw root to treat coughs. Quinault soaked mashed roots in water used to wash swellings and sore eyes. Skagit laid warmed leaves over rheumatic parts, and treated tuberculosis with a root decoction. Lummi drank a root decoction as an emetic. Stuhr says Indians used the plant to treat grippe, eruptions, and consumption. This or some closely-related species has been used in homeopathic and herbal medicine, root as anthelmintic, diuretic, and tonic, flowers, diaphoretic. Gerard recommended it for treating plague and pestilent fevers; Culpeper suggested treating skin spots with the powdered root.

OTHER Western Coltsfoot may be useful in boggy gardens to cover wet waste ground.

Western Coltsfoot
Petasites frigidus
greatly reduced leaf

WESTERN RED COLUMBINE
Aquilegia formosa   Ranunculaceae

NAMES   Columbine refers to the dove, *columba*, in Latin; some think the arched shape of the scarlet petals resembles perched doves. *Aquilegia*, eagle, points out a supposed similarity of the flower spurs to eagle talons. *Formosa* is beautiful in Latin. In June and July the bright red and yellow flowers in somewhat open places attract pollinating butterflies and hummingbirds.

MEDICINE   Quileute chewed Columbine leaves and spit them on burns. Sweet says rubbing mashed, moistened seeds in the hair discourages head lice. Liquid from boiled roots was used as a diarrhea cure; old herbals suggest leaves in a lotion for sore mouths and throats; for jaundice, seeds were steeped in wine. Herbal medicine formerly used some species as diaphoretic, antiscorbutic, and diuretic, but has not done so in recent years. Columbines are not recommended for food. Unlike many other Buttercup family members, they are not known to be poisonous.

OTHER   The design has long been known in heraldry. Columbines can be recognized in illustrated manuscripts from the 15th century. Some Columbines are native to most states of the United States. Many horticultural varieties are available; this native perennial is also desirable for gardens.

SPOTTED or PACIFIC CORAL-ROOT
Corallorhiza maculata   Orchidaceae

NAMES   The Greek *korallion*, coral, and *rhiza*, root, refer to the misshapen roots, which resemble coral growths. *Maculata*, Latin for spotted, describes the lip of this saprophytic or parasitic Orchid, so the common name is a translation of the species. The broad tip of the lower flower petal serves as a landing platform for pollinating insects. The dull-colored flower spikes, eight or more inches tall, are always a pleasant surprise, as they are never common. Coral-root blooms in July under Douglas First or other big trees in second- or old-growth forests.

MEDICINE   Practitioners of herbal and traditional medicine historically have used the root of this perennial or a similar species as diaphoretic, febrifuge, and sedative. Its beauty and need for special growing conditions strongly suggest that it be left undisturbed.

COTTONWOOD
Populus trichocarpa   Salicaceae

NAMES   *Populus*, meaning the common people, was the classical Latin name for Cottonwood, a tree planted in public places in Rome. *Trichocarpa* says hairy-fruited. Pollen-producing and seedbearing catkins borne on separate trees. Its hairy white seeds in late June

Cottonwood
Populus trichocarpa

demonstrate the efficiency of wind distribution. Cottonwood grows rapidly on riverbanks and moist low-lying land.

FOOD   Bella Coola and Northern Kwakiutl ate the fresh cambium layer in May and early June, but did not dry it as it does not keep well. Kirk recommends eating catkins raw or boiled in stew.

MATERIALS   Quinault used Cottonwood for posts around villages. Cowlitz made the hearth board of fire drills from it.   Squaxin used young shoots to make the sweat lodge, also for lashings and for tying thongs.   In recent years, boxes, crates, and pallets have been made from Cottonwood. Toothpicks have been made from *Populus* species trees. Trees planted in 1901 along the Willamette River near Oregon City for paper pulp were largely Cottonwood. It is still considered a good source of paper fiber.

MEDICINE   Squaxin gargled a bark decoction for sore throat, and soaked bruised leaves in water for an antiseptic on cuts.   Klallam prepared an eyewash from Cottonwood buds. Quinault treated cuts and wounds with gum from Cottonwood burls. Traditional and herbal medicine formerly used some other Populus species.

OTHER   Another name for *P. trichocarpa* is the Balsam Poplar. A few buds in February or March, put in hot water, are said to provide a spicy refreshing fragrance. The name Balsam Poplar refers to the Balsam tree ancient Hebrews used to provide sweet smells and oils; it is a member of the Balsamaceae.

COW PARSNIP, MASTERWORT
Heracleum lanatum   Umbelliferae

NAMES   Hercules in the genus refers to size of the plant leaves, while Latin for wooly provides the species. This native North American perennial blooms in April and May.

FOOD   Northwest Indians, including almost every British Columbia group, ate young tops and stems before flowers appeared, often dipping stems in oil, usually peeling and discarding the outer skin. The skin contains a chemical causing sensitization to light, which can also cause blisters. Some people find wet foliage causes dermatitis. Chinook and other native Americans cooked roots, said to taste like rutabaga, and ate the young stems raw. Changing cooking water is desirable. Wildfood authors suggest peeled stems and flower stalks, raw or cooked, as salad; cooked roots; and ashes of dried leaves as a salt substitute. Stems provide a nutritious survival food as they contain about 18 percent protein. Some say they are similar to celery; this information came from books, not personal experience; opinions as to palatability vary!

MEDICINE   Quinault warmed Cow Parsnip leaves to put on sore limbs. Klamath used roots for medicine. California Indians put root pieces in tooth cavities to stop pain and made an infusion for sore throat by soaking the mashed root in water. Early Spaniards treated rheumatism with a root medicine.

Herbal medicine has treated nausea and disorders of stomach and nervous systems with a tea from dried root and seeds and used powdered roots or seeds in a poultice for sore muscles or joints. Cow Parsnip was in the U. S. Pharmacopoeia from 1820 to 1863 as a stimulant and carminative; roots were official, but leaves and fruit were also used; fresh leaves provided a counterirritant.

OTHER   If there is ample boggy space in a home garden, the Cow Parsnip may add interest.

CRABAPPLE
Pyrus fusca (Malus fusca, Pyrus diversifolia)   Rosaceae

NAMES   *Pyrus* is the ancient Latin name for Pear; Pears and Apples are often placed in the same genus. *Fusca*, grayish brown, refers to bark color. The Crabapple which blooms in April over the Old Main trail at Tryon Creek Park is a reminder of past settlers, not the native tree. The tiny Apples make us realize how horticulturalists have increased fruit sizes.

FOOD   Where available, the native Crabapple, more common west of the Cascades, provided an important food for Northwest Indians. Fruit was eaten by all Coastal British Columbia groups and by Lower Thompson, Lower Lillooet and Niska of the Interior. Straits Salish, Halkomelem, Squamish, and Nootka picked Crabapples green and stored them in Cat-tail bags for ripening. They were boiled, steamed in pits, or mashed and mixed with other fruit such as Salal. Swinomish, Samish, and Quileute ate these raw; Quinault and Lower Chinook softened them by storing them in baskets; Cowlitz cooked fruit a little before storing. Niska mixed boiled Crabapples with oil for winter storage. Calapooya stored apples until they turned red, then ate them raw; for winter use, they put them away in oil. Fruit is best for jellies and pickles, as its sharp flavor makes it less desirable raw, though Mohney reports fully ripe Crabapple is palatable. Avoid eating seeds of any Apples; they contain cyanide. Pheasants, grouse, and other birds feed on Crabapples.

MATERIALS Northwest Indians used the tough lightweight wood of the native Crabapple for driving stakes and prongs of spears, often hardened over a fire. Wedges of hard, seasoned wood served to split tree trunks, boards, and firewood, and to hollow out canoes. Lewis and Clark reported use of wedges and axe handles. Quileute, Kwakiutl, Straits Salish, and Halkomelem used it for digging sticks, bows, wedges, and tool handles. Saanich sometimes used Crabapple wood for fishing floats.

MEDICINE Makah made a solution of soaked peeled bark to treat intestinal disorders, dysentery, and diarrhea; Klallam and Quinault used the same infusion as eyewash. Swinomish and Samish used a bark decoction externally to wash cuts, internally for stomach disorders. For lung troubles, Makah chewed leaves which had been soaked in water.

Creeping Charlie
Glecoma hederaceae

CREEPING CHARLIE, GROUND IVY
Glecoma hederaceae (Nepeta glechoma)   Labiatae

NAMES   This low-growing European perennial, blooming from May, grows in moist shaded ground, where its stems take root and creep to nearby areas. No one knows who Charlie was; the Ivy-like growth gave the name Ground Ivy. *Glechon* was the classical Greek name for Pennyroyal, also a Mint family member. The Greek *hedra*, seat, may refer to the way plants cling to walls which it climbs. Like many Mint family members, Creeping Charlie is aromatic. Early Saxons clarified beer with this plant before Hops were introduced; it was in general use until the time of Henry VIII, who reigned from 1509 to 1547.

MEDICINE   Dioscorides, Galen, Parkinson and Gerard recommended this herb; leaves were considered tonic, aperient, diuretic, astringent, stimulant, vulnerary, and good for coughs and headaches. Juice of the fresh herb was also used. A tea is often available at health food stores. Traditional medicine formerly used *G. hederaceae* for pulmonary and urinary ailments, but has not done so for many years.

OTHER   Creeping Charlie is poisonous to stock, especially horses, when eaten in quantity, either fresh or with hay. Adrosko says a 1786 dye list gave *G. hederacea* as a source of the color *Merd'ore'*, goose dung.

RED FLOWERING CURRANT
Ribes sanguineum   Grossulariaceae

NAMES   Pretty red flowers gave this name, also described by *sanguineum*, blood-red in Latin. *Ribes* is from *ribas*, an old Arabic word for a plant with sour juice.

FOOD   Our native Currants all require moist growing conditions; all are edible, but some taste much better; *R. sanguineum* is one of the less palatable species. Washington Klallam, some Oregon and British Columbia groups ate the fresh berries, but did not regard them highly and did not usually dry them. Other groups preferred other Currant species. Some who dislike these berries raw, recommend them for jelly. *R. sanguineum* has also been used to make wine.

OTHER   As Red Flowering Currant is host to White Pine blister rust, various campaigns have eliminated large quantities of this attractive shrub. Seeds of *R. sanguineum* sent to England in 1827 by David Douglas were planted for ornamental use; it still grows in London and Paris parks.

ENGLISH DAISY
Bellis perennis   Compositae

NAMES   The scientific name for the common lawn Daisy, a "day's eye" flower with yellow disk and radiating white ray flowers, comes from Latin, *bellis*, pretty, and *perennis*, as it is a perennial plant.

FOOD   The English Daisy, which has a somewhat acrid taste, has been eaten as a potherb in times of food scarcity. The unpleasant taste makes it unappealing to cattle.

MEDICINE   Gerard said it was good for "all kinds of paines & aches", but herbal medicine has not used it for many years. At the turn of the century traditional and herbal medicine used leaves as a vulnerary and the root as antiscorbutic.

OTHER   This European native is well established west of the Cascades. Its cheerful blooms from early spring to late autumn are favorites with children, who enjoy making Daisy chains. It was often planted in lawns in medieval Europe, but we consider it a weed in ours. Flat leaves close to the ground and a tendency to grow in solid stands help the plant crowd out the more desirable Grasses.

DAISY FLEABANE, HORSEWEED
Conyza (Erigeron) canadensis   Compositae

NAMES   Horses like to eat this weed. The Greek *konyza*, a fly, was used by Pliny and Dioscorides for some Fleabane—*konops* is Greek for flea. Some species with a sticky surface were used in catching flies. *Canadensis* identifies a home of the plant. Fleabanes bloom in August and September, earlier than their close cousins the Asters.

MEDICINE   In a switch from normal, this North American annual is now a widespread pest in the Old World. It had arrived there by the 17th century, for Culpeper and Parkinson said it was astringent, diuretic, and tonic. Traditional medicine formerly had similar uses. Daisy Fleabane causes dermatitis to some people.

DANDELION
Taraxacum officinale   Compositae

NAMES   Dandelion is a corruption of the French *dent de lion*, tooth of a lion, from the jagged leaf shape. *Taraxos*, Greek for disorder, and *akos*, remedy, combine for the genus name; *officinale* confirms medicinal use. Bright yellow flowers and fluffy seedheads of this perennial are most common in early spring and summer, but some bloom continues much of the year. Flowers are sensitive to the intensity of light and close at night or when it rains. The grooved leaves assist maximum use of moisture, as they direct water to the plant center. Dandelions have shiny leaves without hairs, hollow stems with milky juice, and unbranched flower stems with only one flower.

FOOD   Organic Gardening says Dandelions are higher in food values than many cultivated plants. One of the best green vegetable sources of vitamins A and C, Dandelions are also a source of iron, calcium, and potassium. Food use has been recorded from the time of the Roman Empire. This weed, possibly a native of Greece, is thought to have reached America only slightly later than the first colonists. They may have used it for medicine, eaten roots and early spring leaves raw or cooked, roasted roots for a coffee substitute, or used flowers in making wine (yellow blossoms only, the green parts are bitter).

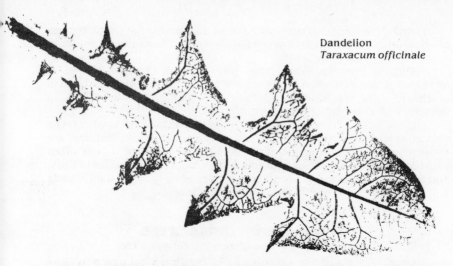

Dandelion
Taraxacum officinale

Changing cooking water will reduce bitterness, as will soaking leaves in salt water for a few hours before using them in salads. Dandelions have been cultivated and harvested like Spinach. The ground dried root can be perked for a coffee substitute, steeped for use as tea, or dried leaves can be used for tea. A Dandelion beer has been made, also a fermented drink made from Dandelion, Stinging Nettles, and Yellow Dock. The plumed seeds make an interesting trail nibble and are recommended as emergency food.

MEDICINE Tillamook, only Northwest group for which Dandelion use was reported, ate cooked greens and made a laxative from the roots. Dandelion has been used for centuries. Medical use was noted in the 10th and 11th centuries by Arabian physicians, in the 13th century in Wales, and by English herbalists in the 16th and 17th centuries. Dandelion was considered most useful combined with other agents; it was an ingredient in many patent medicines. Medicinal value depends on the milky juice. The root, dug in the fall, used fresh or dried, was considered alterative, stomachic, diuretic, laxative, and tonic. It provided a spring tonic used by "our grandmothers." Dandelion was in the U. S. Pharmacopoeia from 1881 to 1926, and the National Formulary until 1965. Most sources no longer consider it desirable. One herbal said Dandelion is overrated as a home remedy; the 1950 U. S. Dispensatory said Dandelion is no longer considered to have therapeutic powers.

OTHER Size of Dandelions varies with growing conditions; in lawns, they grow low; in open spaces they reach up higher. Roots have provided a magenta dye. This is one of the best wildlife food sources, attracting pheasants, Canada geese, grouse, and other birds, small and large forest mammals such as rabbits, elk, deer, bears, and porcupines. It is good forage for domestic stock, and an excellent nectar source.

Pollen of Dandelion is thought to cause hay fever for some; contact dermatitis from touching the flowers has also been reported.

Deer Fern
Blechnum spicant
(l.)fertile frond, sterile frond (r.)

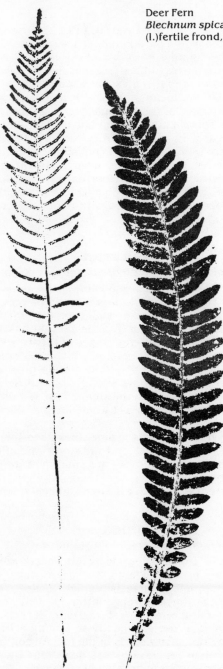

RED DEAD NETTLE
Lamium purpureum   Labiatae

NAMES Reddish-purple flowers and no sting, (Dead), name this plant, which somewhat resembles Stinging Nettle. *Lamium*, from the Greek *laimos*, throat, refers to blossom shape, *purpureum* to foliage color. This invasive European annual with hollow square stems is a nuisance in gardens west of the Cascades, as it spreads efficiently. Its blooms, from May, contain much nectar and are favorites of bees.

FOOD Linnaeus said peasants in Sweden ate *L. purpureum* as a potherb, but current wildfood books do not suggest it. Cattle dislike it.

MEDICINE Herbal medicine has used fresh or dried herb and leaves of Dead Nettle to check hemorrhages, and applied bruised leaves to wounds; the plant was considered astringent and a blood purifier. A dried herb tea, sweetened with honey, was used to induce perspiration and act on the kidneys.

DEER FERN
Blechnum spicant (Struthiopteris spicant)   Polypodiaceae

NAMES Deer Fern provides good forage for domestic and wild animals. Ancient Greeks used *Blechnum* for some fern, probably not this one. The Latin *spica* spike, suggests the tall fertile fronds; spore-bearing leaflets, all more or less the same size, start halfway up the stalk. Deer Fern has two different types of leaves; sterile evergreen fronds form a low crown, in the center of which grow erect, deciduous, spore-bearing stalks. Leaflets of sterile fronds grow almost to the base; those at top and bottom are shorter than those in the middle; leaflets or pinnae are attached to the stem by the full width. Deer Fern grows in heavy shade in moist or wet woods west of the Cascades, and is common in coastal forests.

FOOD Indian hunting parties sometimes used this as field food. Fiddleheads have a nutty flavor and are good raw after the fuzz is removed. As fiddleheads and roots are small and difficult to clean, they are not too practical.

OTHER Deer fern is attractive in shady home gardens.

DOCK
Rumex species   Polygonaceae

NAMES Dock is from the Anglo-Saxon *docce*; *Rumex* from the ancient Latin name used by Pliny. These are mostly troublesome weeds with long taproots. Three perennial Docks: Yellow or Curly Dock, *Rumex crispus*, from Europe; Bitter or Broad-leaved Dock,*R. obtusifolius*, from Europe; and Western Dock, *R. occidentalis*, from western North America, are among those common here. *R. crispus* blooms in June and has showy seeds in the fall. All contain oxalic acid; some taste better than others. In large quantities they are laxative; they are also antiscorbutic. Indians smoked leaves. See Sheep Sorrel for another species.

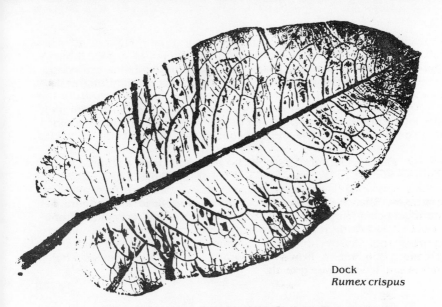

Dock
Rumex crispus

FOOD   Young leaves of several Docks are good raw in salads or boiled, eaten with salt, pepper and lemon juice. Changing cooking water is recommended because of the quantity of acid in the leaves. Curly Dock is preferred. Indians ground the seeds to meal for bread and mush.

MATERIALS   As children, we called the small brown seeds Tobacco; they have been used as a tobacco substitute.

MEDICINE   Some Docks have had medicinal used. *R. obtusifolius* was in the U. S. Pharmacopoeia from 1820 to 1905 for use with skin diseases, as an alterative, laxative and tonic. Roots of *R. crispus,* Curled Dock, used in Europe as laxative, tonic, and alterative, were included in Gerard's Herball. Some medicinal use was also noted for *R. hymenosepalus*, Pie Plant. Docks are among plants suggested to relieve pain from Stinging Nettle; some sources say these are the most effective. There is an old couplet "Nettle in, Dock out, Dock rub the Nettle out."

OTHER   Pima Indians boiled the pounded roots of *R. crispus* to dye cotton yellow; the dry roots were soaked in water for tanning hides. Some Willow wythes used for baskets were dyed yellow with soaked roots; longer soaking produced a brown color.

SPREADING DOGBANE
Apocynum androsaemifolium   Apocynaceae

NAMES   Common and genus names say that this low, branching perennial herb repels dogs; *apo* is Greek for against or away from, and *kyon* means dog. Linnaeus' choice of species name indicates his belief that the leaves resemble those of *Androsace,* a plant of the Primrose family. Dogbane, a serious orchard weed in many areas, grows in somewhat dry, open places, blooming from June. Flowers attract butterflies, bees, flies and moths, and the nectar is a good honey source. It is good that the bitter taste discourages stock browsing, because toxic resins and glycosides in the foliage and milky juice make it dangerous to eat.

MATERIALS Northwest Indians made twine and rope, fishing lures and nets, from Spreading Dogbane fibers. They preferred Black Indian Hemp or Common Dogbane, *A. cannabinum*, when available.

MEDICINE Herbalists formerly used dried Dogbane roots as diuretic, emetic, diaphoretic, tonic, laxative and cardiac stimulant. It had uses similar to Digitalis, but was considered inferior. *Apocynum* species may have dangerous side effects; some also have allergic responses. Rhizome and roots of *A. androsaemifolium* were in the U. S. Pharmacopoeia from 1820 to 1882 and the National Formulary from 1942 to 1950 as diuretic, cathartic, emetic, and expectorant. Vogel reported that North American Indians had used boiled roots of Spreading Dogbane to provide temporary sterility. A root tea as a hair rinse is said to stimulate growth.

DOGFENNEL, MAYWEED
Anthemis cotula   Compositae

NAMES An unpleasant smell and finely-cut leaves, resembling those of Fennel, mark Dogfennel, also called Stinking Chamomile. *Anthemis* is the ancient Greek name of Chamomile, another genus member. *Cotula,* Latin for small cup, refers to shape of flower heads. This annual Eurasian herb grows freely in waste places, blooming from May until late summer.

MEDICINE The whole herb was an accredited drug in 1897, with tonic, antispasmodic, emmenagogue, and emetic uses reported. As medication, flowers are said to be slightly less disagreeable than leaves. Contact with leaves and flowers causes dermatitis for some, and its acrid juice may cause blisters, but rubbing leaves of *A. cotula* on bee stings has been suggested. Dogfennel has been used to drive away fleas.

OTHER Adrosko reported Dogfennel as a dye source in 1786; color given was Ronce' d' Artois; *ronce* means briar in French, but the town name Artois does not help identify the color.

CREEK or RED OSIER DOGWOOD
Cornus stolonifera (C. sericea)   Cornaceae

NAMES Creek identifies habitat and Red Osier describes color of the pliable twigs of this shrub. The name Dogwood could come from the fact that mangy dogs were formerly washed with this species. Latin *stolonifera* reports runners that root. Its small white May flowers are less attractive than those of the Dogwood tree, which are enhanced by showy white bracts. Leaves turn red in the fall.

68

FOOD Kootenay and Green River chewed berry kernels. Thompson, Lillooet, Okanagan, Shuswap, and Kootenay ate the bitter berries, usually fresh, often pounded and mixed with Service Berries, which ripen later. Lower Lillooet mashed berries and saved the stones, which they stored and ate like peanuts. Sturtevant reported the bitter and unpalatable fruit was eaten by Indians of the Missouri River. Some British Columbia Indians and those of eastern North America used leaves and inner bark of Creek Dogwood as a substitute for smoking tobacco, or mixed this with tobacco or other dried leaves. Creek Dogwood provides popular browse for deer, and fruits eaten by birds and animals. The edible white fruits contain enough protein, fat, and carbohydrate to be valuable wildlife food. This sometimes results in overbrowsing.

MATERIALS From the thin, flexible branches of Creek Dogwood, Shuswap made salmon stretchers, Bella Coola made barbecue racks, Haida, drying racks and other frames. Some used Creek Dogwood for bows, others, for crotches of slingshots. Oregon Indians used sticks to hold salmon open during the smoking process, also used *C. stolonifera* and Willow to make baby baskets. These slipper-shaped baskets, carried on the mother's back, contained a little seat and sometimes a small inverted basket as a sunshade.

Okanagan and Shuswap made cordage for tying and lashing from bark fiber. The bark was stripped, twisted, and spliced, or the entire branch was twisted. This cord was used to tie fish traps, barbecue sticks, smokehouse poles and latticework for fish weirs. Wood was used to smoke and dry meat and fish as it doesn't blacken the meat. Creek Dogwood was sometimes combined with Service Berry wood for fuel to dry Huckleberries.

MEDICINE Snohomish treated sore eyes with a solution of boiled scraped bark. British Columbia Carrier smoked bark for lung sickness. Oregon Indians used a tea from inner bark of Creek Dogwood as a quinine substitute. Some herbal medicine uses for *C. stolonifera* are the same as those for *C. nuttallii*.

OTHER Creek Dogwood provides attractive foliage, flowers, fruit, and valuable wildlife cover in a wild garden. This native shrub helps rehabilitate moist habitats.

PACIFIC or WESTERN FLOWERING DOGWOOD
Cornus nuttallii   Cornaceae

NAMES Some sources say that skewers or "dags" were made from this wood, hence dagwood or Dogwood. Latin for horn, *cornu*, refers to hardness of the wood. The name of this handsome tree, provincial flower of British Columbia, protected there by law, honors the botanist Nuttall. White petals of the flowers, blooming in April, are really modified bracts and the true flowers are the inconspicuous center portion of what we admire. The red, fall-ripening fruits are eaten by birds, squirrels and chipmunks. It is never abundant. Foliage often turns deep red in the autumn.

FOOD Sturtevant lists eight *Cornus* species, with somewhat edible fruit or berries, not including *C. nuttallii* or *C. florida*. *C. mas* of Europe and Asia Minor, the Cornelian cherry, was formerly cultivated for its fruit, although reports of its palatability are not all favorable. It grows at Portland's Leach Botanical Park and other local gardens.

MATERIALS Scraped inner bark and dried roasted leaves of Pacific Dogwood have been added to smoking tobacco. Green River, Skagit, and Klallam used Dogwood disks for the gambling game. Snohomish used Dogwood pounding sticks to separate fibers from roasted Brake Fern roots. The hard wood is useful for tool handles and wedges.

MEDICINE Dogwood bark has been used as a quinine substitute. Lummi and Green River boiled peeled bark for a laxative. Boiling dried root or bark of this or Creek Dogwood provides a mildly stimulative cold and fever remedy; chewing twigs of *Cornus* species is said to ward off fever. During the Civil War, Florida Dogwood was used to treat malaria; natives of eastern North America had long used *Cornus* species for fevers. The National Formulary formerly included dried root and bark of *C. florida* as astringent, tonic, febrifuge; they were considered stimulant and emetic when fresh. Western Dogwood is believed to have similar properties. In early days, Dogwood twigs were used for toothbrushes. In eastern North America a bark decoction from *C. florida* was used to treat sore mouths.

OTHER David Douglas, who discovered Western Dogwood, believed it to be the Florida Dogwood; Nuttall recognized it as a different species. *C. florida*, (the name means flowering freely), normally grows up to 20 or 30 feet tall; its white bracts are indented at the tip. *C. nuttallii*, largest tree in the Dogwood family, may grow 60 feet tall and reach 150 years of age. Trees with trunks from 6 to 12 inches through are estimated as from 50 to 100 years old.

Dogwood is a good garden ornamental; it sometimes blooms again in late summer, but is attractive at all seasons. Flowers of scattered trees are an effective contrast with adjacent trees in April. Florida Dogwood, also desirable for gardens, is in plantings at Portland's Japanese Garden.

DOUGLAS FIR
Pseudotsuga menziesii (P. taxifolia)   Pinaceae

NAMES Menzies discovered Douglas Fir on Vancouver Island in 1791 while serving in the British navy; Douglas' explorations in 1825-27 are recognized in the common name. Seeds he sent to England grew well, so the tree was introduced there at an early period in our history. Explorers thought this a Fir, a tree in the genus *Abies*. The Greek term *pseudo* means false, though *tsuga* is Japanese for Hemlock, not Fir. Many know this by the previously-accepted name, *P. taxifolia*, which translates as Yew-leaved False Hemlock. The present name was published in 1950; others date from 1803. Of the five species in the genus *Pseudotsuga*, one grows in California and three in Eastern Asia and Japan.

The soft needles of Douglas Fir grow around the stem like a bottle brush; the lower surface of the leaves or needles is whitish, composed of two rows of stomata separated by a midvein. It retains these evergreen needles from 7 to 10 years. Douglas Fir has small, short-lived pollen cones. Fruit is a cone, which matures at the end of the first season and falls entire. These distinctive woody seed cones, which hang down from the branches, have three-part bracts. These "pitchfork" bracts extend beyond the cone scales. Douglas Fir cones are always the same because there is only one species in our area. Most people identify almost any cones as "Pine" cones, which are far less common west of the Cascades than Douglas Fir cones. Pine cones come in many sizes and types, depending on the species. Neither Pine nor true Fir cones have the three-part bracts, and Fir cones shatter so are not found whole.

FOOD Klallam, Cowlitz, and Quinault chewed Douglas Fir pitch like gum. Both whites and native Americans have made a beverage tea, rich in vitamin C, from young Douglas Fir needles. Shuswap occasionally ate the small, pitchy seeds.

MATERIALS Douglas Fir, the Oregon state tree, is the most common evergreen tree in our area, a very important contributor to Oregon's economy, and the most important lumber tree in the nation. In the 1950s it provided more than 20 percent of saw timber volume in the United States. It occurs in even-aged stands or mixtures. This fast-growing tree is processed into lumber, sliced for veneer, chipped for paper, ground for particle board. Where available, Douglas Fir is the leading Christmas tree. Trees planted in the Tillamook burn of the 1930s are now being harvested, a time factor showing its desirability as a timber crop. Mature trees may be 100 or more feet tall when 125 to 200 years old; its life span may reach 300 or more years. The thick, deeply-furrowed bark provides protection from fire and temperature extremes.

Douglas Fir was an important firewood for most Northwest Washington and coastal British Columbia groups, bark being especially desirable. Quinault, Bella Coola and Lillooet made torches of pitchy parts. It was not used for woodwork as it does not split evenly, but was used for harpoon shafts and salmon spears by Quinault, Cowlitz, Skagit, Lummi, Klallam, Swinomish, Chehalis and Green River. Quinault also used it for handles of dip nets. Turpentine made from Douglas Fir has been marketed under the name of Oregon Balsam. In the past, Douglas Fir pitch was gathered abundantly; it was processed into paints, soap, India ink, and used for medical purposes.

MEDICINE Cowlitz, Quinault and Skagit used Douglas Fir pitch on sores. Cowlitz and Squaxin boiled pitch for cold medicine and made tea for colds with Douglas Fir needles mixed with Cedar. Swinomish boiled needles for a general tonic, boiled bark of young roots and drank the liquid for colds, or used it to bathe babies. Swinomish chewed bud tips for sores in the mouth or sore throats; Skagit boiled bark for an antiseptic; Thompson made a tonic tea from young twigs and needles; Kootenay chewed dried sap as a gum to treat colds. Stuhr says Indians used Douglas Fir in sweat baths to cure rheumatism.

OTHER To keep hunters' scent from deer and insure successful hunting, some Northern California bow and arrow hunters passed through smoke of Douglas Fir boughs before hunting; use of proper incantations was important.

Douglas Fir needs a great deal of light and is called an intolerant tree, one which cannot grow in the shade. A climax forest is considered one in which the trees maintain a stable population, reproducing themselves as they die off or are destroyed. Since Douglas Fir seedlings need more light than is available in the shade, it cannot reproduce in a forest, hence is not a part of a climax forest. Douglas Fir may live to be so old that its presence or absence in such a forest is not obvious to us. . .we tend not to think in terms of hundreds of years!

Swinomish boiled bark for a light brown dye to make nets invisible to fish. Some Indian groups split smaller roots of Douglas Fir and other conifers for basket making.

DUCKFOOT, INSIDE-OUT-FLOWER
Vancouveria hexandra Berberidaceae

NAMES Duckfoot pictures leaf shape, helping recognition when flowers are not available; Inside-out-flower describes the curiously-reversed arrangement of flower parts. *Vancouveria* honors Captain George Vancouver, who explored Puget Sound and lower Columbia River areas between 1790 and 1795. *Hexandra,* Greek for six, says floral parts are in sixes.

OTHER This May-blooming perennial native is an attractive groundcover for a woodland garden. There are only three species of *Vancouveria,* all in coastal North America; the others grow in Southern Oregon and Northern California. All are related to the Oregon Grape.

BLUE ELDERBERRY
Sambucus cerulea (S. glauca) Caprifoliaceae

NAMES Starting fires by blowing air through hollow stems of Elder is recognized by *aeld* Anglo-Saxon for fire. A similar plant was called *Sambucus* in the time of Pliny. The sambuke is an ancient Greek stringed instrument; some say it was made with Elder wood. *Cerulea* refers to the blue berries; Blue Elder grows on both sides of the Cascades.

FOOD Flat-topped white flower clusters in open areas and woodland edges in June and July are followed by frosty-blue berries in late summer. Elderberry flowers are a nice addition to pancakes; flower heads dipped in batter make good fritters. Blue Elderberries are reported excellent raw or cooked, for preserves, pies, hot drinks, as a substitute for raisins, for wine. Flavor is considered better after the first frost.

Elderberries have long been used in Europe, species in different areas having similar uses. Berries are high in fat, protein, and carbohydrates, but seeds of some species are toxic containing prussic (hydrocyanic) acid. Some sources recommend straining out seeds to eliminate toxins when using fruit for jams and wine. Coast Salish ate Blue Elderberries where available, though distribution was spotty. California Indians ate Blue Elderberries fresh and dried, and also used them for a beverage. Klallam, Chehalis, Squaxin and other Northwest Washington Indians ate the blue as they did the red. Skokomish ate the blue fresh. Blue Elderberries, fresh, cooked, or dried, were a favorite food of Calapooya of Western Oregon.

MATERIALS Indians dried Elderberry stems before use, a time-consuming process during which poisons dissipate. Northwest Indians used Elderberry for arrow shafts and made wind instruments from dried stems after poking out the pith with hot sticks. To California Indians the Elder was the Tree of Music; they used it to make flutes. Some Northern California tribes made dance clappers and twirling sticks for fire by friction from Elder. Lummi harvested Elderberries with a comb made of Syringa.

Present opinion advises against using Elderberry for blowguns, whistles, and other craft items due to poisonous alkaloids and cyanogenic glycosides. These poisons exist in greater quantities in leaves, stems, and roots than in the unripe berries and flowers; ripening or cooking berries renders the poisons harmless. Ripe Blue Elderberries can be safely eaten raw, but best advice seems to suggest cooking Red Elderberries.

MEDICINE Klallam used a tea of steeped bark for diarrhea, Quinault a bark tea for an emetic. Paiute and Shoshone boiled roots of Red or Blue Elderberries until soft and applied these to inflammations. A syrup of Blue Elderberries has been an official medicine in England to treat coughs and colds. European herbals included medicinal uses for flowers, juice, and roots of Elders. Fresh flowers have provided an antiseptic wash for skin diseases, and a decoction of fresh leaves used to check bleeding of the lungs. Inner bark made a strong emetic. Juice of crushed leaves is an antidote for Nettle stings.

Berries and flowers of an eastern *Sambucus* were in the U. S. Pharmacopoeia from 1831 to 1905, included in the National Formulary in 1916 and 1917. Elder Flower Water, used in perfume or flavoring, has also been official at various times. A current herbal suggests a tea from Blue Elderberry flowers to break fevers and a diuretic from the leaves. A burn ointment has been made from Elderberry juice and grease. Elderberry bark and roots are no longer considered desirable for medicine.

OTHER Songbirds, robins, bandtailed pigeons and grouse enjoy these berries; rabbits, squirrels, chipmunks, mice, and rats eat fruit and bark of both Red and Blue Elder; deer eat foliage and twigs. A decoction of Elderberry leaves is said to keep caterpillars from eating garden plants. The National Geographic of February, 1919, reported the odor of Sweet Elder repelled insects, and an infusion of Elder leaves kept bugs from vines. An 18th century gardener recommended whipping growing vegetables with young Elder twigs for protection from insect damage.

Red Elderberry
Sambucus racemosa

Elderberry species are native in temperate zones throughout the world, growing in southern South America, Eurasia, Europe, Australia, tropical Asia, and North America. English country folk believed an Elderberry bush was safe during a storm because the Cross was made from Elder and even lightning would not strike it. Conflicting lore says that Elder is witchridden and no witch of any standing would be without an Elder in her garden, also that before cutting down an Elder, one must apologize to the witch inside to avoid dire consequences.

Blue Elderberries have been used to dye wool; bark and root produce a black dye; berries mordanted with alum and salt, a lilac-blue or purple dye, and leaves with an alum mordant provide a green dye.

RED ELDERBERRY
Sambucus racemosa (S. callicarpa)   Caprifoliaceae

NAMES *Racemosa* describes the flower head: a raceme. Botanists currently identify both European and American forms of the Red Elderberry as *S. racemosa*. There are more species of Blue than of Red Elderberries, so many references included only Blue or Black Elderberries. Gunther reported about Red Elderberry, which grows west of the Cascades and was more widely available to groups she studied. Red Elderberry flowers, a half-ball mound of small white blooms, appear in March and April, well before the flat-topped flower clusters of Blue Elderberry.

FOOD Widespread in coastal British Columbia, Washington, and Oregon, Red Elderberries were extensively used for food by Coastal groups and are still utilized, although not highly regarded and often mixed with Blueberries, Salal, or other berries to add palatability. Chehalis, Cowlitz, Green River, Klallam, Makah, Quileute, Quinault, Skagit, Skokomish, Skykomish, Snohomish, Squaxin, Swinomish, and Lower Chinook used Red Elderberries. Berries, cooked in steaming pits or cedar boxes, were dried on a rack over a small fire for 24 hours to make cakes. Sometimes Skunk Cabbage leaves were used to wrap these for winter storage. With the recent addition of sugar, berries are now used for jam, jelly, and wine. Kingsbury warns that bark, wood, leaves, and roots of Elderberry are poisonous, and says that uncooked red berries may produce nausea, but are not usually fatal. Children have been poisoned by eating unripe berries. Number of leaflets in the compound leaves vary in different species; Red Elderberry has mostly five to seven, Blue, seven to nine or eleven.

Comments about edibility of Red Elderberry have varied in Oregon State Extension Service Bulletin 697. The 1966 edition stated Red Elderberries were not edible. The 1975 revision said Red Elderberries, edible though not tasty, were good for jelly, jam, juice drinks, and wine.

Some sources suggest avoiding Red Elderberries; an acquaintance was outraged at the suggestion that Red Elderberries were poisonous. She thinks them better than the blue and always uses them. Wildfood books which list the Red Elderberry as edible fairly consistently say the blue berries seem more palatable to more people.

MEDICINE   Makah put pounded fresh leaves of Red Elderberry on abscesses or boils; Cowlitz used ground leaves on sore joints to reduce swelling. Squaxin mashed leaves, dipped the pulp into hot water and applied this to areas infected with blood poisoning. Other medicinal uses are noted under Blue Elderberry. Watch for Elderberry bushes if Stinging Nettles are common—you may need some mashed leaves.

OTHER   Blue and Red Elderberry are desirable for wild gardens, though they may need heavy pruning. The red berries are bright spots during early summer. Red Elderberry likes some shade and is happy in wooded areas, while Blue prefers the open.   Both attract birds; chipmunks and squirrels like them, too.

Enchanter's Nightshade
Circaea alpina

ENCHANTER'S NIGHTSHADE
Circaea alpina (C. pacifica)   Onagraceae

NAMES   Night is often associated with sorcery; Circe was a sorceress in classic Greek mythology. As plants of this genus grow in damp, cool areas supposedly attractive to witches, choice of genus and common names has logic. *Alpina* describes one habitat, to which it is not limited.

OTHER   This invasive native of moist woodlands, blooming in May and June, is a weed we didn't import from Europe!

EVENING PRIMROSE
Oenothera species   Primulaceae

NAMES   These fragrant flowers open about dusk, conveniently for night-flying insects upon which they depend for pollination. This explains the common name, though flowers also open in the daytime, especially after some early blooms have been pollinated. The Greek *oinos* wine, and *thera*, herb, provided the old Greek name for some plant eaten to promote a relish for wines; relationship with this genus is unknown. *O. biennis*, Common Evening Primrose, is a biennial or short-lived perennial wildflower; gone wild is the cultivated hybrid *O. erythrosepala*, the Red-sepaled Evening Primrose. Both grow mostly west of the Cascades, and bloom from June.

FOOD   Wildfood authors suggest roots cooked, and young shoots as salad or potherbs, admitting this may be an acquired taste. *O. biennis*, a North American Evening Primrose, was introduced and cultivated for food in Europe as early as 1614.

MEDICINE   Traditional and herbal medicine have used *O. biennis* as an antispasmodic for coughs, and for treating skin irritations. Other herbal medicine uses were as astringent, diuretic, and sedative, and for a skin ointment. Cough syrup has been prepared from the roots or tops with honey. The stems contain tannin.

OTHER   Both *O. biennis* and *O. erythrosepala* are sun-loving plants desirable for wild gardens. Evening Primroses are not related to Primroses, genus *Primula*, family Primulaceae.

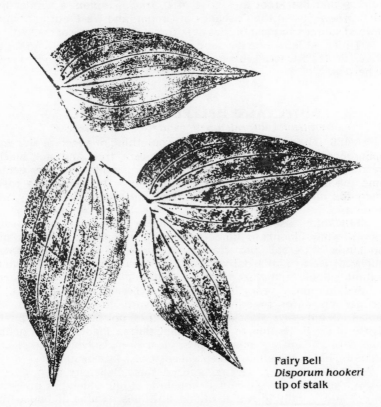

Fairy Bell
Disporum hookeri
tip of stalk

FAIRY BELL and FAIRY LANTERN
Disporum hookeri (D. oreganum), D. smithii   Liliaceae

NAMES   Greek *dis*, double, and *spora*, seed, say some species have two ovules in the cells. Both species honor botanists. Fairy Bell is appropriate for *D. hookeri*, *(D. oreganum)*, as stamens hang below the petals of the small blooms like a bell clapper. Fairy Lantern fits the larger flower of *D. smithii*, which has stamens shorter than the petals. Calling this Smith's Fairy Bells honors a celebrated English botanist, Dr. J. E. Smith, 1759-1828.

These white-belled perennials bloom in moist woodlands about April, often as neighbors. Curved stems and leaves with "drip tips" help drain off heavy rainfall. It is hard to tell which is which when they are not blooming.   Stems of *D. hookeri* are said to be sparingly-branched, those of *D. smithii*, freely-branched. *D. hookeri* is more common in the areas I know. The bright red to orange-red berries do not help, as both go through similar color phases.

FOOD   Fruit of Fairy Bells was not highly regarded by most British Columbia groups, but Thompson ate raw berries of *D. hookeri.* Shuswap and Blackfoot ate those of *D. trachycarpum*, a similar lily which grows east of the Cascades in Oregon and Washington. Some wildfood sources suggest berries of *D. trachycarpum*, raw or cooked.

OTHER   At least one Portland area nursery handles *D. smithii; D. hookeri* is also attractive in a home garden. Wild birds and rodents eat the berries.

WHITE FALSE HELLEBORE, CORN LILY
Veratrum californicum (V. caudatum)   Liliaceae

NAMES   The name Corn Lily gives something of a feel for size and appearance of this tall leafy plant. Latin *vera*, true, and *ater*, black, referring to black roots of some Hellebore, combine for this genus name, used in ancient times for Hellebore.   The Christmas rose, *Helleborus niger*, grown in gardens here, is a true Hellebore. See Names, p.

MEDICINE   False Hellebore, a coarse perennial plant with greenish-white blooms in June, grows in or near moist meadows and woodlands.   Herbal and traditional medicine have used False Hellebores as arterial sedatives, irritants, emetics, and as a nervous excitant. Roots and rhizome of *V. californicum* and *V. viride* were included for many years in the National Formulary and the U. S. Pharmacopoeia for lowering blood pressure; an overdose causes nausea. Dennis says most reported *Veratrum* poisonings were due to misuse of medicine, not to eating plant parts. Plants of this genus cause dermatitis in some individuals. Most animals avoid eating False Hellebore, but occasionally sheep will eat it with fatal results.

Northwest Indians, who recognized False Hellebore as a violent poison, cautiously used it and its cousin *V. viride*, which has greenish flowers and usually grows at higher altitudes.   Quinault boiled the whole plant and administered small doses of the liquid to treat

False Hellebore
Veratrum californicum
greatly reduced leaf

rheumatism. British Columbia groups also had medicinal uses for the False Hellebores. Some Nevada Indians used a root tea as a contraceptive, fresh roots for temporary and the dried root for permanent sterility. Shoshone treated snakebite with the crushed raw root, and chewed the raw root for sore throats and colds.

OTHER Parkinson reported Spaniards used juice from roots of a European *Veratrum* to poison arrows and daggers. Powdered roots of *V. californicum* are an effective insecticide, but its use has been limited as it is toxic to mammals.

FALSE SOLOMON'S SEAL; STAR-FLOWERED FALSE SOLO-MON'S SEAL
Smilacina racemosa (S. amplexicaulis); S. stellata   Liliaceae

NAMES An eastern plant, *Polygonatum multiflorum*, was called Solomon's Seal because of a supposed resemblance of root scars to seals, or because of the plant's ability to "seal" wounds. False says these plants look alike, and the plants are similar, though the eastern plant has white flowers at leaf axils rather than at the end of the leaf stalk. *Smilacina* is a diminutive of the plant name *Smilax;* this reference is not clear. *Racemosa* is logical: the flower cluster is a terminal raceme.

In mid-April fluffy white flowers top the curved stem of False Solomon's Seal, *S. racemosa.* A few starry flowers provide both common and species names of the similar but smaller Star-flowered False Solomon's Seal, *S. stellata*, (Latin, *stella*, star.)

FOOD Though not especially palatable, berries of both these perennials were sometimes eaten, raw or cooked, by Northwest Indians, who also ate young shoots as a potherb and rootstocks parboiled after soaking them with wood ashes to remove bitterness. Wildfood editors suggest the same uses. Ruffed grouse enjoy the red-spotted berries.

MEDICINE Berries of *S. racemosa* are cathartic if eaten raw in quantity, less so if cooked. Paiute and Shoshone used False Solomon's Seal leaves in a contraceptive tea and treated female trouble and internal pain with a root tea. Roots of *S. stellata* were collected in the fall, sliced, then dried and strung for later use to treat bleeding wounds. Herbalists have used dried roots of both species as demulcent and expectorant, to treat lung infections and sore throat.

OTHER Both False Solomon's Seal and Star-Flowered False Solomon's Seal make attractive additions to wild gardens. Many gardeners in our area also enjoy Solomon's Seal, *Polygonatum multiflorum.*

FAWN LILY, DEERTONGUE
Erythronium oregonum   Liliaceae

NAMES Imaginative common names of *Erythronium* are shared by several different species and also include Dogtooth Violet, Lambstongue, Adder's Tongue, and Trout Lily. *Erythronium* comes from the Greek *erythro*, red; some species have red flowers. This pretty native perennial blooms in late March; its white flower petals have orange at the base. Most *Erythroniums* grow in mountain areas; *E. oregonum* is an exception, found in Oak and open coniferous woods of valleys and low hills. It thrives in cultivation; keep pet rabbits away if you wish to enjoy it another year.

FOOD   Leaves and bulbs of some alpine species of *Erythronium* were eaten by Northwest Indians. *E. oregonum* should not be eaten. It is said to poison poultry which eat it.

CALIFORNIA FIGWORT
Scrophularia californica   Scrophulariaceae

NAMES   Common and species names suggest Figwort is common in California, but it also grows freely at the Oregon coast and other Northwest sites west of the Cascades. Fig describes root shape of some species. *Scrophularia* came from use of plants of the genus to treat scrofula. Another name, Bee Plant, tells us why beekeepers sometimes cultivate Figwort.

MEDICINE   Herbalists prepare a tincture, tea, or ointment from the above-ground plant for treating skin eruptions and infections, as an alterative, and a blood purifier. Species members contain cardioactive glycosides.

OTHER   This native perennial adds interest to moist sunny places in large wild gardens. The maroon-colored flowers are small relative to the coarse green foliage, three or more feet tall.

FILAREE, STORK'S BILL
Erodium cicutarium   Geraniaceae

NAMES   The sharp beak of the conspicuously-shaped seed or fruit gave common and genus names: *alfillerilla*, Spanish for needle, and Stork's Bill; *erodium*, from the Greek *erodios*, heron. *Cicutarium* suggests leaves resemble those of Water Hemlock, genus *Cicuta*. This European annual weed blooms on dry open ground much of the year.

FOOD and MEDICINE   Filaree, picked before flowering, is eaten for salad greens or as a potherb. Herbal and traditional medicine formerly used it as astringent and diuretic.

OTHER   Filaree provides valuable forage for livestock, especially sheep. Plant size varies with growing conditions.

FIREWEED, WILLOW HERB
Epilobium angustifolium   Onagraceae

NAMES   Common names are descriptive; the bright purple-pink flowers give burned-over areas a flame-like appearance; some Willows have narrow leaves. *Angustifolium* combines Latin for narrow and leaf. The scientific name is informative: *epi* is Greek for upon, *lobos*, for pod; flowers are above the developing seed pods.

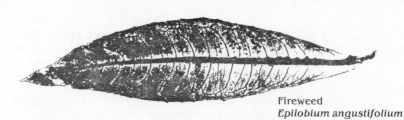

Fireweed
Epilobium angustifolium

Fireweed, widely distributed in the northern hemisphere, is a pioneer plant, among the first to take over after fire or clearing. Survival of such plants depends upon fast reproduction.

FOOD   Northwest Indians and other peoples throughout the world ate Fireweed sprouts raw or cooked, also sometimes ate roots of these perennials.   Wildfood books still recommend these uses.   Sechelt, Squamish, Bella Coola, Haida, and Tsimshian ate the inner part of the stem.   Indians made a beverage tea, rich in vitamin C, from dried Fireweed leaves; health food stores often sell a similar tea.   Fireweed honey is a classic treat.   Pith of Fireweed has been used to thicken stews and soups, and dried pith used to brew ale.   Domestic livestock and wild animals relish the plant.

MATERIALS   Puget Sound Indians used fluff from Fireweed seeds to extend their supply of fibers from dogs and mountain goats. Quinault and Skokomish added it to duck feathers.   Haida split Fireweed stems, pulled them through their teeth to eat the pith, and saved the fibers to make twine for fish nets.

MEDICINE   Haida ate Fireweed shoots for a laxative.   A root decoction was used by Snohomish to treat sore throat, by Skokomish for tuberculosis.   Swinomish bathed invalids in a liquid prepared by boiling the whole plant.   Herbal and traditional medicine have used Fireweed leaves and roots as demulcent, tonic, and astringent.

OTHER   Some sources recommend Fireweed for home gardens; it seems too aggressive! Adrosko called *E. angustifolium* French willow and said the leaves and flowers provided a dye color called "vigogna"; this is the name of a town in Spain known for the woolen hats made there in the latter 18th century.

Many *Epilobium* species grow in our area.   The Tall Annual Willow Herb, *E. paniculatum*, is a garden weed. Others have attractive flowers, many pinkish, some yellow; several species are alpine.   All have been used for beverage tea or as greens and potherbs early in the season.

FOAMFLOWER, COOLWORT, FALSE MITREWORT
Tiarella trifoliata   Saxifragaceae

NAMES   White flowers of this perennial look like whipped foam; three-part plant leaves are also recognized by the species name.   Cool and wort, from the Anglo-Saxon *wyrt*, plant or herb, fit Foamflower's home in shady moist woods, where it blooms over an extended period from late April.   *Tiarella*, diminutive of the Greek *tiara*, an ancient

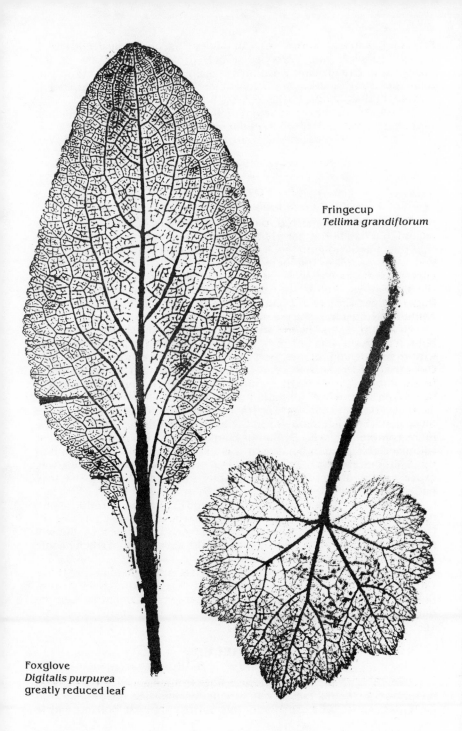

Fringecup
Tellima grandiflorum

Foxglove
Digitalis purpurea
greatly reduced leaf

Persian headdress, refers to fruit shape, as does False Mitrewort. *Tiarella trifoliata* var. *trifoliata,* which grows mostly west of the Cascades in Oregon and Washington, seems more common than var. *unifoliata.* Variety *laciniata* is more limited in its range; one place it grows is in the Olympic National Park. See page

MEDICINE   Quileute   chewed   leaves   as   a   cough   medicine. Herbalists have used this or a similar species as tonic and diuretic.

OTHER   Foamflower makes an attractive ground cover in a wild garden.

FOXGLOVE
Digitalis purpurea   Scrophulariaceae

NAMES   One suggested source of the name Foxglove is a European legend that bad fairies gave these flowers to foxes to cover their feet so that they might go more quietly to the henhouse! Another says Fox is derived from Folk's, folk being good fairies. *Digit* is Latin for finger; coverings are noted in these common names: Foxglove, Goblins' Gloves, Fairies', Folks', and Witches' Gloves. Fuchs recognized the finger-shaped flowers of *D. purpurea* by the German name for thimble, *Fingerhut,* finger hat. *Purpurea* is for flower color.See p. .

MEDICINE   This heart stimulant, one of the most important botanical drugs, was in the U. S. Pharmacopoeia for many years. As with many biennial plants used for medicine, leaves were gathered in the spring of the second year, before plant strength was diverted to flower production.   Small quantities strengthen and stimulate the heartbeat, improving circulation of the muscle. It has also been used as a diuretic. It is safe only in small doses as it contains a poisonous glycoside which accumulates in the body. A synthetic product is now more commonly used. Benjamin Franklin was an early heart patient who used this drug in London; it was used in England in 1785 and in the United States by 1787. There is a story that when *Digitalis* was experimental, a doctor's wife urged him to prescribe this for charity patients only!

OTHER   Northwest Indians recognized Foxglove as introduced and developed no uses. It is well established in the wild, especially west of the Cascades. Stock have been poisoned by eating the leaves, but will not eat it if other forage is available. Deaths have resulted when people mistakenly used Foxglove leaves for tea.

This is a handsome garden plant. It should be avoided by those with children too young to understand that sucking nectar from flowers or eating leaves or seeds of garden ornamentals can be dangerous.

FRINGECUP
Tellima grandiflorum   Saxifragaceae

NAMES   The cup-shaped flowers have fringed edges. *Tellima* is an anagram of *Mitella,* small cap, name of another *Saxifrage,* used perhaps because seed capsules of this native perennial look like a little bishop's mitre. *Grandiflorum* is easy to decipher, but hard to understand, as the flowers are not large. Stems of pinkish-white flowers rise above the green leaves in April. As many *Saxifrage* have similar leaves, species identification may be difficult when flowers are not blooming.

MEDICINE Skagit pounded the Fringecup plant, boiled it, and drank the resultant liquid to treat sickness, especially to help restore appetite.

Garlic Mustard
Alliaria officinalis

GARLIC MUSTARD
Alliaria officinalis (Sisymbrium alliaria)   Cruciferae

NAMES Crushed leaves have a definite garlic scent, mentioned in the common name; the genus name comes from the Latin *allium,* onion. Garlic Mustard's white flowers appear from early April; four-petaled flowers and long seedpods are typical of the Mustard family.

FOOD As a potherb, Garlic Mustard should be picked as bloom approaches. The garlic flavor suggests use in salads or sauces, where Grieve says it warms the stomach and strengthens digestive faculties. Euell Gibbons recommended the leaves on meat sandwiches. Sturtevant said it was fried with bacon in Wales. Cows which eat this plant produce milk with a disagreeable flavor.

MEDICINE Herbal and traditional medicine formerly used herb and seeds of Garlic Mustard as antiscorbutic, diuretic, diaphoretic, and expectorant. A volatile oil similar to Mustard oil has been extracted from the root.

OTHER In 1964, Garlic Mustard was established in the Northwest only in the Portland area. This European biennial weed is crowding out native plants, as do many introductions which lack the enemies of their homelands. Its explosive reproduction at Tryon Creek Park has approached geometric rather than arithmetic progression.

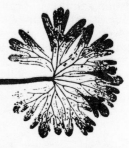

GERANIUM, CRANE'S BILL
Geranium species   Geraniaceae

NAMES   Because of the long seed beak, *geranos,* Greek for crane, is appropriate for genus and common names. *Geranium dissectum,* cut-leaf Geranium, and *G. molle,* Dovefoot Geranium, annual European weeds, may be those you see blooming from March or April and recognize as Geraniums by the distinctive seed pods. The Latin *dissectum* confirms the Cut Leaf of the common name; *molle,* Latin for soft, describes the soft petals of the Dovefoot Geranium. These are relatively common, but there are others; comparing different species helps in identifying Geraniums.

FOOD   Food uses do not seem to be listed, though I remember jelly flavored with Geranium leaves; perhaps that was a cultivated variety?

MEDICINE   Chehalis recognized *G. molle* as adventitious and did not use it.   Native Americans and settlers in other areas used Geraniums when astringents were needed; it contains tannic and gallic acids. Some Indians used a Geranium for birth control. Roots and leaves of *Geranium* species, including *G. dissectum,* were formerly used in traditional and herbal medicine. The dried rhizome of *G. maculatum,* native of Canada and eastern North America, was in the U. S. Pharmacopoeia from 1820 to 1916 and the National Formulary from 1916 to 1936 as tonic, astringent and styptic. It was used for sore gums, pyorrhea, and for toothaches. The tannin was the effective part, which also made it useful in tanning leather.

GOATSBEARD
Aruncus sylvester   Rosaceae

NAMES   The Greek *aryngos,* goatsbeard, gives the common name, and *sylvester,* woodland, identifies habitat.   Tall stalks of this Northwest native perennial are topped in May by fluffy white flowers, perhaps beardlike!

MEDICINE   Skagit and Klallam mixed ashes of Goatsbeard roots and twigs with bear grease to treat swellings and sores; Quileute treated sores with scraped root pulp. Quileute boiled the pounded root for a general tonic also used to treat colds and sore throat, while Skagit and Makah used the same decoction for kidney trouble.

OTHER Goatsbeard is an herb, as the stem is not woody and dies back each year, although its vigorous growth and large size give it a shrub-like appearance. Male and female flowers grow on separate plants. The showier male flower clusters are suggested for moist areas of wild gardens.

GOLDENROD
Solidago species   Compositae

NAMES Golden rod describes the flower spike. *Solidago*, Latin meaning to make solid or draw together, refers to use of Goldenrod to treat wounds. Whether the species we see blooming from July is *S. canadensis, (S. elongata), S. occidentalis*, or another relative, we all recognize these perennial herbs and have heard that Goldenrod causes hay fever. Scientists now say Goldenrod does not cause hay fever, as the pollen is so heavy that the wind cannot carry it great distances.

FOOD Wildfood sources suggest young leaves of some species as a potherb and leaves and flowers of any species for a beverage tea. Euell Gibbons in Stalking the Wild Asparagus doesn't mention potherbs, and limits his tea to one eastern species , *S. odora.*

MEDICINE California Indians used a decoction of Goldenrod leaves to heal open sores and to wash wounds of men or horses; powdered dried leaves were also used to treat sores. Herbalists recognized astringent and antiseptic properties of Goldenrods by using them to treat wounds and ulcers. A lotion to treat sores can be made from leaves and stems of any *Solidago.* Grieve listed *S. canadensis* for vulnerary use. Traditional medicine included *S. odora* in the U. S. Pharmacopoeia in 1850 as aromatic, diuretic, and astringent.

OTHER There are nearly 100 Goldenrods, mostly native to North America, so identification can be a botanist's nightmare; Hitchcock lists eleven species in our area.

Leaves of many Goldenrods contain rubber in varying amounts. Fresh or dried flowers, gathered when coming into bloom, make a yellow dye. Eight to twelve quarts of flowers are needed to dye one pound of wool. Goldenrod, surely an attractive flower, has been suggested for wild gardens, but may not be considered because it is so common.

GRAND FIR, WHITE FIR
Abies grandis   Pinaceae

NAMES *Abies* is the Latin name for Fir; common and species names mean the same. This is the only true Fir found at sea level in Oregon; as it seldom grows in pure stands, commercial use has been limited. Under favorable conditions, Grand Fir may reach 75 to 125 feet at maturity, may grow more than 250 feet tall, and have a trunk two to three feet in diameter; it may live for 200 to 250 years. Its needles, arranged in two flattened rows, have blunt rather than pointed ends. Cones of true Firs stand erect on the branches and shatter at maturity, so cone scales rather than complete cones will be found below Fir trees. Most true Firs in the Pacific Northwest grow in the mountains.

MATERIALS   Chehalis of Washington and some British Columbia groups used Grand Fir for fuel. British Columbia groups finished canoe paddles and wooden articles with the pitch. Nootka, Thompson, and others used the boughs as bedding, floor coverings, and berry-basket covers. Kwakiutl used branches for ritual scrubbing in purification rites. Grand Fir is used for paper pulp.

MEDICINE   Green River boiled needles for a cold remedy. Samish, Swinomish and Skagit pounded Grand Fir with the Stinging Nettle plant and boiled this for a general tonic or a cold medicine. Vogel says Indian and folk medicine used the resins of Balsam Fir, *A. balsama*. Pitch of some *Abies* species has been used to scent creams, oils, salves, soaps, and shampoos. Stuhr says decoctions of *A. concolor*, White Fir, were used as a diuretic and to treat malaria.

OTHER   Flathead of Montana pulverized Grand Fir needles for baby powder and body scent. Straits Salish of Vancouver Island made a brown dye for basketry from Grand Fir bark, and a pink dye from bark and red ochre.

COMMON GROUNDSEL, RAGWORT
Senecio vulgaris   Compositae

NAMES   Groundsel, from the Anglo-Saxon *grundeswelge*, ground swallower, says the plant spreads rapidly. The name rag plant, (wort), suits its appearance. *Senecio*, from *senex*, Latin for an old man, refers to hairlike white bristles on the seeds, or to the bald seedhead after seeds are gone. *Vulgaris* means common. This annual European weed grows in disturbed soil and waste places west of the Cascades.

Ragwort, reasonably easy to pull in the garden, is prolific and keeps coming back. There are some 1,000 species of Groundsels or Ragworts in the genus *Senecio*, growing mostly in temperate regions; Hitchcock lists more than 30 in Flora of the Pacific Northwest.

MEDICINE   Herbalists have used dried herb and fresh juice as diaphoretic, antiscorbutic, diuretic, purgative, and anthelmintic. Boiling water poured over the fresh plant makes an old-fashioned remedy for chapped hands. *S. aureus* of Canada and eastern United States was an ingredient in Lydia Pinkham's Vegetable Compound; this species is said to be the least toxic *Senecio*, as most are toxic.

OTHER   Leaves and seeds provide food for canaries and other wild species; rabbits like Groundsel. Common Groundsel, related to Tansy Ragwort, is also poisonous to stock, cattle and horses being most susceptible.

HARDHACK, STEEPLE BUSH
Spiraea douglasii   Rosaceae

NAMES Hardhack states a fact: it is hard to hack a way through dense growth of this shrub, which prefers moist or marshy areas. Pink bloom spikes in June or July, reaching up like a steeple, or dried-up remains after pink has faded to brown, make Hardhack easy to identify. Some species have twisted seedpods, reported by the Greek *speiran*, to twist. The species name, established in 1833, honors David Douglas, Scots botanist and plant explorer. Douglas reported finding it on the "North-West coast of America, about the Columbia and the Straits of de Fuca."

MATERIALS Bella Coola made halibut hooks from fire-hardened wood, also hooks for drying and smoking salmon. Vancouver Island Salish made blades, halibut hooks, and scrapers from Hardhack. Nootka made a tool for gathering dentalium shells; Northwest peoples used these shells, sometimes called wampum, as currency before white traders came. Lummi used Hardhack stems for roasting salmon, and Quinault used peeled stems to string clams for roasting. These are good for roasting hotdogs.

MEDICINE Lummi treated diarrhea with a tea from Hardhack seeds. Herbal medicine used root and leaves of some *Spiraea* as astringent and tonic. An eastern *Spiraea* was formerly in the U. S. Pharmacopoeia, flowers as a diuretic, roots as an astringent for diarrhea. The "spirin" of aspirin came from *Spiraea*, as salicylic acid was first isolated from a *Spiraea*.

OTHER At least one wild plant nursery near Portland stocks this shrub. Non-native *Spiraeas* are less aggressive and may be more satisfactory in home gardens.

HAWKBIT, FALL DANDELION; HAIRY HAWKBIT
Leontodon autumnalis; L. nudicaulis   Compositae

NAMES There is an ancient belief that hawks ate Hawkbit or Hawkweed plants to sharpen their eyesight. Dandelion and *Leontodon* say the same thing in different languages. Dandelion is from the French, while *Leontodon* combines the Greek *leon*, lion, with *odous*, tooth; both refer to tooth-like edges of plant leaves. *L. autumnalis* has several flower heads at the top of its flower stalk. *L. nudicaulis* has unbranched flower stems.

MEDICINE Medicinal uses of these Eurasian perennials are similar to those of Dandelion. Herbal medicine used Hawkbits to treat jaundice and as a diuretic.

OTHER As with other "Dandelions," these weeds are not always easy to identify. They perpetuate themselves so successfully that there are lots available to study! Knowing whether the plant juice is milky helps identification, as does whether leaves are smooth or hairy, stems single or branched. All have yellow blooms. Sometimes flowering time is helpful; *L. autumnalis*, as the species name infers, is chiefly fall-blooming. *Nudicaulis* says the stem is smooth, which seems inappropriate for the Hairy Hawkbit.

ROUGH HAWKSBEARD; SMOOTH HAWKSBEARD
Crepis setosa; C. capillaris   Compositae

NAMES *Crepis,* from the Greek *krepis,* a sandal, says the deeply-cut leaves reminded someone of open sandals. *Setose* means full of bristles, appropriate for the Rough Hawksbeard. *Capillaris* says hairlike; this species has short hairs on the leaves and lower stem. Bristly or Rough Hawksbeard is an annual; the Smooth Hawksbeard is an annual, occasionally biennial.

OTHER These two European imports, blooming from May to November, are found west of the Cascades. Both Hawksbeards and Hawkbits were included by Linnaeus in 1753.

BLACK HAWTHORN; RED or ENGLISH HAWTHORN
Crataegus douglasii; C. oxyacantha   Rosaceae

NAMES Haw, from the Anglo-Saxon *haga,* hedge, is another name for fruit such as this and Rose hips; anyone who has grabbed a branch of this small tree or large shrub will know about the thorn. Red and Black describe fruit colors. The Greek *kratos,* strength, refers to the strong tough wood of Hawthorns. The Northwest native bears the name of David Douglas, credited with collecting an early specimen. *Oxus* means sharp, and *akantha,* a thorn.

FOOD Kirk notes five species of *Crataegus* with edible berries, good raw or for jams and jellies, better after first frost. Adding peels of lemon, orange, or grapefruit makes an interesting marmalade. Sturtevant reported English Hawthorn berries had been eaten by Scottish Highlanders, but were seldom eaten in England except by children. Native Americans ate berries fresh, dried, or mixed in pemmican. Several British Columbia groups ate the dry, seedy, fruits of native Black Hawthorn, sometimes boiled in Cedar boxes and stored for winter, but considered them poor quality. Bella Coola now use them for jam. Oregon Indians sometimes mixed Hawthorn fruit with other berries in dried cakes. Leaves have been used as an adulterant for tea. Roasted seeds of *Crataegus* species have provided a coffee substitute.

MEDICINE Hawthorn was used in herbal medicine, fruits as astringent, tonic, diuretic, and cardiac, flowers and berries for a decoction to treat sore throat.

OTHER The perfumed flowers are said to attract carrion insect pollinators. A folk belief is that Hawthorn flowers still smell of the Great Plague of London. Lore and legend go back before the time of the druids. Paul Bunyan was said to use a Hawthorn back scratcher.

The native Black Hawthorn, blooming in April or May, is not valued for gardens, though it provides good cover for birds and small animals. Birds have spread English Hawthorn from ornamental plantings; it grows more often west of the Cascades, while *C. douglasii* grows on both sides.

Adrosko reported English Hawthorn was the source of a dye, color given as mordore.

English Hawthorn
Crataegus douglasii
tip of branch

Hazelnut
Corylus cornuta

WESTERN HAZEL
Corylus cornuta   Betulaceae

NAMES Western Hazel, wild relative of the Filbert, grows mainly in woods and thickets west of the Cascades. *Corylus* is from the Greek *korys*, a helmet, for the wrapper enclosing the nuts or fruits, which often grow in pairs. *Cornuta*, horned, also refers to the fruit covering. Its February flowers appear before leaves, aiding pollination.

FOOD Known edibility of this nut and its relations dates from archaeological records of Mesolithic man in Europe. Forest growth of the period shows that their animals browsed on Hazel.

Hazelnuts were important food for native Americans and are popular with squirrels, chipmunks, mice, bluejays, songbirds, crows, Clark's nutcracker and other wildlife. Cowlitz, Squaxin, Chehalis, and Puyallup ate these fresh and stored some for winter. Most British Columbia groups picked Hazelnuts early and allowed them to ripen before eating them. Skagit, Lummi, Snohomish, and Swinomish ate them fresh but did not store them. Some groups roasted them in coals.

Wildfood editors are enthusiastic about Hazelnuts, with good reason, but try and find them when ripe! A fatty oil from the nut has been used to adulterate Almond oil or as a food oil in its own right. The ground nut is delicious as a flour in breads, cakes, and puddings. Ingredients of many gourmet recipes, both European and domestic, include Hazelnuts.

MATERIALS Chehalis used twisted Hazel twigs for ties, but not for ropes; Skokomish twisted long twigs to make rope. Some groups used Hazel wythes for lashing house or smokehouse frames. Northern California Indians used split stems of this shrub or small tree for their finest baskets. Hoopa used Hazel for sieves, some fine enough for flour; Wailaki used Hazel for sifters; Pomo made baby baskets from Hazel wood. Pioneers sometimes made the bows of ox yokes from Hazel.

Those looking for hotdog roasting sticks in the woods should know Western Hazel, which is safe for this use. It has several stems from one base, many close to the right size, so plenty will remain to support plant life if some are picked.

MEDICINE The U. S. Dispensatory reported a vermifuge was formerly made from scaly coverings of the Hazelnut; because of its disagreeable nature, it has fallen from use. The sharp spicules of the nut coverings were scraped off and mixed with syrup or molasses; these penetrated bodies of round worms in the digestive system and killed them.

OTHER Pendant male catkins before the leaves unfold in early spring make Western Hazel attractive for a home garden; the female flowers are small and inconspicuous. Pollen of Hazel species is one cause of hay fever.

Heal-all
Prunella vulgaris

Hedge Nettle
Stachys cooleyae

93

HEAL-ALL, SELF- or ALL-HEAL
Prunella vulgaris Labiatae

NAMES Common names reflect the high regard in which this European native, blooming in moist places from June, was formerly held. Linnaeus established the name *Prunella*, a variant of *Brunella*, German for sore throat, for which this was considered a cure. *Vulgaris* is Latin for common. Specialists recognize an American plant, not sufficiently different to justify a separate species, by use of a variety. This perennial lawn pest adapts to mowing by reducing its size while creeping rootstocks continue its spread. Heal-all usually grows larger in wild surroundings. The flower shape is adapted for pollination by bees.

The Doctrine of Signatures, an antiquated theory that plant shape identifies medicinal use, gave the name Carpenter Weed. Some carpenters' tools possibly had a similar shape, so the plant was considered suitable to treat ailments of carpenters! Proponents of this theory recommended red plants for blood problems, plants with heart-shaped leaves to treat heart ailments.

FOOD Crushed leaves, fresh or dried, can be used for a beverage tea.

MEDICINE Quinault and Quileute treated boils with Heal-all. Gerard recommended it for treating "inward and outward" wounds. Herbal uses were as astringent, antispasmodic, tonic and styptic. A tea of the dried plant was used as a gargle for sore throat. The fresh plant served as an antiseptic poultice for bruises and scrapes; crushed leaves and flower spikes treated bites and scratches. Heal-all is high in tannin content. Many herbal medicines were alcohol based; some suspect that their virtues did not all come from the plant content.

OTHER A 1786 dye list said Heal-all gives the color olive.

HEDGE NETTLE or WOUNDWORT
Stachys cooleyae Labiatae

NAMES Hedge Nettle describes growth of this native, which blooms from June; *stachys*, Greek for spike, identifies the flower pattern. Botanist Grace E. Cooley provided the type specimen in 1891 and the species name uses the feminine form, *ae*. Many Mint family plants have a scent which is released by crushing or bruising the leaves; that of Hedge Nettle is less pleasant than some others. Like Dead Nettle, also in the Mint family, Hedge Nettle has no sting. Stinging Nettle belongs to the Nettle family, *Urticaceae*.

FOOD and MATERIALS Wildfood editors suggest eating the tubers in the fall, raw or cooked. Makah and Quinault used the plant to cover steaming sprouts.

MEDICINE Green River treated boils, and Puyallup and Saanich made a spring tonic with crushed roots. Quileute used Hedge Nettle leaves in a steam bath. Woundwort, also used for other Stachys species, notes uses recommended by Gerard, Parkinson, and other early herbalists: washes or poultices of fresh leaves to treat sores and wounds.

Western Hemlock
Tsuga heterophylla
branch tip, small piece
shows different needle lengths

WESTERN HEMLOCK
Tsuga heterophylla   Pinaceae

NAMES Hemlock comes from the Anglo-Saxon *hemlic* or *hymlic*. *Tsuga* is Japanese for Hemlock. The Greek *heterophylla*, different leaves, points out varying needle lengths; all are shorter than Douglas Fir needles. Hemlock, Washington's state tree, grows well under shady conditions and is especially abundant in rainy and foggy areas. Its small cones are 3/4 to an inch long. Its drooping top or "leader" helps identification.

FOOD Northwestern Washington Indians did not use Hemlock for food, though Cowlitz used branch tips to flavor bear meat. Haida, Tsimshian, Bella Coola, Kwakiutl, and Vancouver Island Samish, British Columbia Coastal groups, and Niska of the Interior, ate fresh cambium or dried it in cakes for winter use. Leaves have been used for a beverage tea.

MATERIALS Washington Indians used Hemlock in many ways. They made wedges to hollow out canoes and split boards from seasoned limbs and knots, soaked bark to line cooking pits; Quileute stored Elderberries in Hemlock bark containers; it provided firewood. British Columbia groups found Hemlock a useful wood, moderately heavy, durable, and reasonably easy to work. It was used for totem poles, canoes, carved spoons, roasting spits, combs, dipnet poles, Elderberry-picking hooks. Haida and Straits Salish made Halibut and cod hooks from rotted Hemlock stumps. Haida carved large feast dishes from Hemlock stumps. Hemlock gum, like that of other conifers, was sometimes used as a glue.

Hemlock is an important source of paper pulp in the Pacific Northwest. It is also used for lumber and plywood.

MEDICINE Cowlitz used pitch, Quinault pitch and ground bark, for a face paint which helped prevent chapped skin. Quinault made a laxative, Cowlitz and Skagit a decoction for sore eyes or skin sores, from boiled bark, examples of differing uses of a similar solution by different groups. Bark, sometimes combined with Licorice Fern, provided a tea to treat hemorrhages. Chewed young Hemlock tips were spit on swellings to reduce them. Pitch was reported helpful for healing sores. Some northern tribes used tea from Hemlock leaves and twigs for colds, tuberculosis, bronchitis and pleurisy, and poultices of chewed bark for boils. Nootka used the bark as an astringent to stop external bleeding.

OTHER High tannin content of bark made it useful for dyeing wool and basketry materials. Klallam, Lummi, and Makah boiled pounded inner bark in salt water for a red paint and wood preservative for spears and paddles. Kwakiutl steeped bark in urine for a black dye. Bark also provided a red-brown dye used in tanning hides.

Hemlock trees may reach a height of 100 to 200 feet, grow 2 to 6 feet in diameter, and live from two to five hundred years. Western Hemlock and Mountain Hemlock, *T. mertensiana*, are excellent ornamentals.

ENGLISH HOLLY
Ilex aquifolium   Aquifoliaceae

NAMES Holly comes from the Anglo-Saxon *holen* or *holeyn*. *Ilex* may be a variant of the Celtic *ac* or *ec*, point, as the leaves are spiny. One possible source of the species name is the Latin *acus*, needle, and *folium*, leaf; another is *aqua*, water, as the shiny evergreen leaves may appear wet. The leaves decay slowly after falling. Male and female flowers are on different plants. Birds introduced this European shrub into wild areas after eating berries from ornamental plantings.

FOOD Leaves of all *Ilex* species contain caffeine; Holly leaves have been used as a tea substitute.

MEDICINE Herbals formerly recommended leaves as diaphoretic, tonic, antiarthritic, and as a treatment for rheumatism. Avoid eating berries, which are diuretic, violently emetic, and purgative—dangerous but not fatally. Indians in North Carolina and Florida used a native Holly, *I. cassine*, to make a so-called black drink or yaupon. A strong leaf solution was used as an emetic; weaker solutions provided a diuretic and a stimulant beverage.

OTHER Traditional use of Holly as a Christmas decoration dates from Roman Christians; it was used earlier in pagan rites. Shoots and leaves were listed in a 1786 dye list as *ventre de crapaud*, toad's belly; modern dyers have obtained a gray-green color.

HONESTY, SILVER DOLLAR, MONEY PLANT
Lunaria annua   Cruciferae

NAMES Seedpod shape provides common and genus names: Silver Dollar, Money Plant, and Honesty refer to the round pods, as does *Lunaria*, Latin for moon. Latin for year, *annua*, says this European native is an annual.

FOOD and MEDICINE Young spring shoots of Money Plant are used in green salads or cooked as a potherb. Flowers are said to have a tonic effect.

OTHER Honesty's pinky-purple blooms come in April, but this garden escape is valued for the spectacular seedpods, popular for winter bouquets.

ORANGE HONEYSUCKLE
Lonicera ciliosa   Caprifoliaceae

NAMES Nectar can be sucked from these orange flowers. Linnaeus chose Lonicera to honor a 16th century German botanist, Adam Lonitzer. *Ciliosa* means hairlike in Latin; leaves have marginal hairs. A pair of joined leaves make a cup just beneath the flower cluster.

FOOD Some Oregon Indians ate the berries, which were nowhere abundant. Though Honeysuckle was not a food source for British Columbia groups, their children enjoyed the nectar. Present day wildfood editors recommend berries of *L. ciliosa*, *L. involucrata*, and *L.*

utahensis, raw, dried, or for jams and jellies. These natives have not been reported poisonous, but some *Lonicera* have toxic berries, toxicity varying with the species. See Twinberry.

MEDICINE Swinomish chewed Orange Honeysuckle leaves and swallowed the juice to treat colds, and made a bark decoction for colds and sore throat. Lummi treated tuberculosis with a drink from boiled leaves. Klallam put chewed leaves on bruises. Chehalis and Squaxin bathed little girls with water from crushed leaves, believing it would make their hair long and sleek. Flathead boiled Honeysuckle stems for a shampoo said to make hair grow longer. Thompson combined fiber from Honeysuckle stems with other fibers to weave mats, bags, caps, aprons, and blankets.

OTHER This native vine, blooming in April in open woods and thickets should be used more often in wild gardens. Hummingbirds and Swallowtail butterflies pollinate the Honeysuckle; bees like the nectar.

COMMON or FIELD HORSETAIL
Equisetum arvense   Equisetaceae

NAMES See p. 27. Horsetails reproduce by spores, do not have flowers, and lack leaves of a familiar type. They are sometimes listed as Fern allies. The number of ridges on stems of Horsetail are identifying features. *E. arvense* has 10 to 12, the larger *E. telmateia* has 20 to 40. *E. telmateia*, Giant Horsetail, grows west of the Cascades in our range. It and Marsh Horsetail, *E. palustre,* are less common here than *E. arvense.*

FOOD Common Horsetail has two different life forms; the familiar branched form is sterile and lasts through one growing season. Spore-bearing whitish shoots appear briefly in the spring and soon die back. Several Coast Salish groups, Lower Chinook, and some Oregon Indians ate the fertile shoots of the Common Horsetail; Klallam, Makah, Quileute, and Quinault ate reproductive shoots of the Giant Horsetail. Shoots were eaten raw or cooked; tough outer fibers were peeled off or chewed and discarded. Some wildfood books currently suggest young reproductive shoots, peeled and boiled. Green stems of the sterile cycle, ground to a powder, have been used for thickening and for a beverage tea. Some native groups also ate roots, tubers, and peeled stems, raw or cooked; this is not recommended at the present time. Sweet says 17th century Romans boiled Horsetail heads or dipped them in flour and fried them.

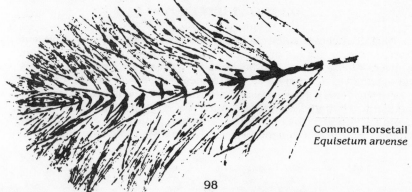

Common Horsetail
Equisetum arvense

98

MEDICINE Horsetails contain aconitic acid. Some Nevada Indians used ashes of *E. arvense* to treat skin and mouth sores. Quinault used root preparations from Giant Horsetail as a wash for sore eyes. Makah ate heads of reproductive shoots of *E. telmateia* raw as a diarrhea cure. Cowlitz boiled broken stalks to provide a wash for hair infested with lice. Cheyenne treated horses with coughs with an infusion of stems and leaves of the Common Horsetail. Herbal medicine has used the stems to treat kidney disorders, as a diuretic, astringent, emmenagogue, and vulnerary. Gerard, Galen and Dioscorides recommended Horsetail as a vulnerary. Fresh sterile stems, cut off just above the roots, were considered most effective. Traditional medicine has used Horsetail as an astringent and diuretic.

OTHER Swinomish used *E. telmateia* for scouring and for polishing arrow shafts. Jointed stems of Horsetail explode when the plant is thrown into a fire, a discovery some medicine men used to stimulate their patients. Shoshone used *E. arvense* to make whistles. Horsetail is poisonous to horses, especially in hay, because it causes a vitamin B1 deficiency. It apparently causes fewer problems for cows or goats.

Common Horsetail is a frustratingly durable remnant from the time of the dinosaurs, when Equisetum species were a far greater part of the plant population. Horsetail, some as large as our present trees, decayed to form much of our coal. Ability to grow through asphalt and persistent return in gardens demonstrate the strength of the perennial rootstocks, which spread underground. Presence of Horsetail indicates available water. See also Scouring Rush.

RED HUCKLEBERRY
Vaccinium parvifolium   Ericaceae

NAMES Huckleberry may be a corruption of whortleberry, from the Anglo-Saxon *wyrtil*, or small wort, plus berry. *Vaccinium* from Latin *vacca*, cow, was the ancient name for some berried plant browsed by cattle. *Parvifolium* describes the leaves, which may or may not be smaller than those of most Huckleberries; (Latin, *parvus*, small). Red Huckleberry, typical of moist coniferous forests, often grows from tree stumps, courtesy of the birds. This deciduous shrub, with ridged and angled green stems, grows west of the Cascades. Foliage is browsed by blacktailed deer; berries are a favorite food of birds, squirrels, chipmunks, bears—and people!

FOOD Western Washington, Oregon, and British Columbia groups fortunate enough to have Red Huckleberries ate them fresh or dried. Red Huckleberries were used to flavor stews. Berries were brushed or combed from the bushes. Quinault used Red Huckleberry leaves for tea; leaves of other *Vaccinium* species have also been used for beverage teas; palatability varies in different species. Both red berries and the blue-black berries of the Coast Huckleberry, *V. ovatum*, are excellent for jams, jellies, pies, and syrup. Wildfood editors also recommend drying these; because of high moisture content, Red Huckleberries take longer to dry than some other fruits.

Red Huckleberry
Vaccinium parvifolium

Indian Plum
Oemleria cerasiformis

MEDICINE   Skagit boiled bark of *V. parvifolium* for a cold remedy. Herbal medicine has used Huckleberry leaves as astringent and diuretic, and as a wash for skin irritations. Folk medicine has treated diabetes with some *Vaccinium* species.

OTHER   Huckleberries are good ornamentals for wild gardens, though plants may be hard to find in nurseries. Both *V. parvifolium* and *V. ovatum* will grow in open areas or partial shade. The evergreen leaves of Coast Huckleberry are attractive all year long; florists harvest this handsome foliage, shipping some to eastern markets.

INDIAN PIPE
Monotropa uniflora   Ericaceae

NAMES   Indian Pipe pictures the drooping pipe-like flowers. *Mono* or one, plus *tropos*, turned, describes the white flowers, one to a stem, turned in one direction. This saprophytic plant with its roots associated with fungus grows in decaying vegetation, often under Douglas Firs. Indian Pipe is becoming more rare as the shaded woods of its favored habitat become less available. If you are fortunate enough to find this interesting plant, leave it to revisit another season. Its clear white does not survive picking and will turn black. Its needs are very specific; because it appears one year does not mean it will return the next.

FOOD and MEDICINE   Kirk says Indian Pipe is edible, raw or cooked. Some American Indians used juice of Indian Pipe with water as an eye lotion, though this was not reported in Northwest sources. Herbal medicine has used roots of Indian Pipe as tonic, sedative and antispasmodic, and for an eyewash.

INDIAN PLUM, OSO BERRY
Oemleria (Osmaronia) cerasiformis   Rosaceae

NAMES   *Oso* means bear in Spanish. The species name says cherry-formed, but the fruit is more plum-shaped, as noted in the common name. *Oemleria* honors a European botanist. The former genus name combines *osme*, Greek for smell, or fragrant, and *aronia*, a generic name for some Rose family member. Rank scent of the blooms attracts pollinating insects. To be aware of how common this shrub is, come to Northwest woods in late February and see flowers and leaves unfolding well ahead of other plants. The female (pistillate), flowers are often in full bloom as the leaves appear. Male and female flowers grow on different plants; the pollen-bearing male flower is larger. Indian Plum leaves are among the first to yellow and drop to the ground. This genus has only one species; it grows mostly west of the Cascades.

FOOD   The small fruit which ripens in early summer is not too tasty, though it was eaten by Samish, Swinomish, Chehalis, Squaxin, Quinault, Skagit, Lummi, Snohomish, British Columbia and other groups, some of whom considered it starvation food, some, walking snacks. The seed is large in proportion to fruit size. Cowlitz dried it for winter use. British Columbia groups ate Indian Plum with oil and stored it for winter in Cedar boxes.

Mohney says growing conditions influence size and quantity more in Indian Plum than in any other wild fruit. The cucumber-like flavor of young leaves makes them desirable for salads; they become bitter as they grow older. Wildfood authors recommend fruit raw or cooked; cooking reduces the bitter flavor. Indian Plum, a favorite of birds and small rodents, seldom hangs on the branches after ripening.

OTHER   This fast-growing shrub, often associated with Alder and Bigleaf Maple, is helpful in land reclamation.

OREGON IRIS
Iris tenax   Iridaceae

NAMES   The name of Iris, classical Greek goddess symbolized by the rainbow, is appropriate for a genus which has flowers of many colors. April flowers of *I. tenax* vary from white to deep purples.

MATERIALS   Leaves of this native perennial, described by the Latin *tenax*, strong or tough, provided an important fiber source to Northwest Indians. The highly-valued cords used for twine, ropes, and snares took much time and patience to make. Oregon Iris prefers open areas west of the Cascades; its presence indicates water close to the surface.

MEDICINE   Oregon Iris, *I. tenax*, had some herbal medicine use, but *I. versicolor* was considered best for medicine. Roots of *I. versicolor*, which grows in the eastern United States were in the U. S. Pharmacopoeia from 1820 to 1895 and the National Formulary from 1916 to 1942; roots of all *Iris* species are said to be emetic and cathartic. Herbal medicine has treated vomiting and depression with a tincture of *Iris* plants. Oregon Iris is poisonous to stock, the irritant largely contained in leaves and rootstocks. It sometimes causes contact dermatitis.

OTHER   Oregon Iris is attractive in a rock garden. It grows in well-drained soil in partial shade to full sun. *Iris, fleur de lys,* is the heraldic emblem of France.

ENGLISH IVY
Hedera helix   Araliaceae

NAMES   Ivy is a variant of the Anglo-Saxon *ifig*. The Celtic *hedra*, a cord, describes the stems, while the Greek *helix*, to turn round, points out twining stem growth. This evergreen European perennial with shiny leaves has been scattered by birds who enjoyed the berries. Perhaps the best-known vine in cultivation, this invasive creeper travels by putting down small roots along its stem; these rootlets may be aerial or reach directly into the ground. It provides food and shelter to birds and late fall nectar for bees.

MEDICINE   Fresh Ivy leaves were formerly used for local applications in skin diseases; they contain the glycoside hederin. Berries are emetic, diaphoretic, stimulant and cathartic. Gum of Ivy has been used to relieve toothache. Ivy is no longer highly regarded for medicine. Berries and leaves are poisonous when eaten, causing illness but not

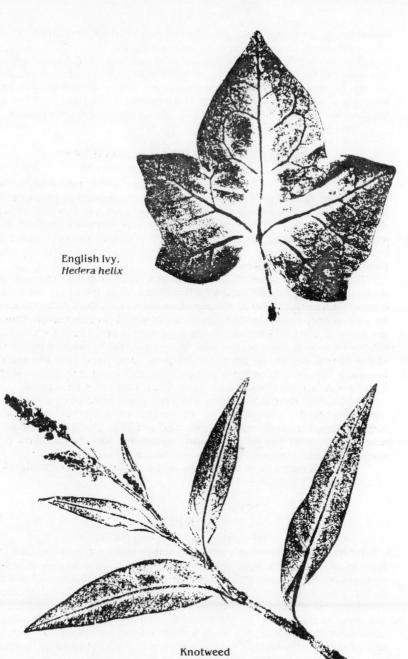

English Ivy,
Hedera helix

Knotweed
Polygonum aviculare
tip of stalk

fatalities in children. Some find Ivy causes contact dermatitis. Cattle which ate leaf clippings of ivy have been poisoned.

OTHER   English Ivy, widely used as an ornamental, has escaped to the wild in Western Oregon and Washington. Uncontrolled wild Ivy causes serious problems by climbing trees, where its weight and bulk increase wind resistance and help cause storm-felled trees. A 1786 dye list gave wood of Ivy as a source of the color ombre or brownish yellow.

KNOTWEED, DOORWEED, KNOTGRASS
Polygonum species   Polygonaceae

NAMES   Greek *poly* means many; *gonum*, knee, refers to swollen nodes on knotlike stems of some species, a description appropriate for Knotweed and Knotgrass. It is said that *P. aviculare*, an annual European weed which grows freely in waste spaces, has followed civilized man wherever he has gone. Seeds have been eaten whole, and ground for use as flour. *Aviculare* is the diminutive of the Latin *avis* or bird; many small birds eat the seeds of this species.

FOOD   Conspicuous along roadsides near Seaside, Oregon, are great masses of the tall Japanese Knotweed, *P. cuspidatum*, an escape from ornamental plantings. It has hollow stems and spreads by rhizomes as well as seeds. Its branched panicles of small greenish-white flowers are attractive in late summer. It also grew wild in areas about which Euell Gibbons wrote in Stalking the Wild Asparagus. He recommended peeled stalks of this aggressive perennial for pies, sauce, and jam; the sprouts he used for salads and potherbs, also roots; all are quite acidic. *Cuspidatum* is from the Latin for pointed, or terminating in a point.

MEDICINE   Herbal medicine has used fresh juice of P. aviculare to stop nosebleeds, and seeds as an emetic or purgative; roots were formerly used as vulnerary, styptic and diuretic. Traditional medicine also used *P. aviculare* as astringent, vulnerary, and styptic. Bark from the rhizome of *P. cuspidatum* has been used as a laxative.

OTHER   Leaves of *P. aviculare* make an indigo dye; roots of *P. cuspidatum*, a yellow dye.

LADY FERN
Athyrium filix-femina   Polypodiaceae

NAMES   The common name translates the species: *filix*, Latin for fern, plus *femina*, woman. Greek for without, *a*, plus *thyros*, a shield, refers to the lack of spore covering in this fern, a technicality not obvious to the casual observer. Lady Fern's leafy circle grows in moist areas, often six feet or more tall. Fronds are said to be like a lady, small at the top and bottom, big in the middle! Leaves or fronds are shaped somewhat like parentheses, with leaflets narrow at top and bottom and widest at the center. Still differently stated, the frond tapers fairly evenly from the center to each end; leaflets or pinnae grow quite close to the stem base. Many people confuse Lady and Spreading Wood Fern; Wood Fern fronds are roughly triangular and the leaflets start farther from the base, giving the frond a longer stem.

Lady Fern
Athyrium filix-femina
tip of frond showing spore pattern

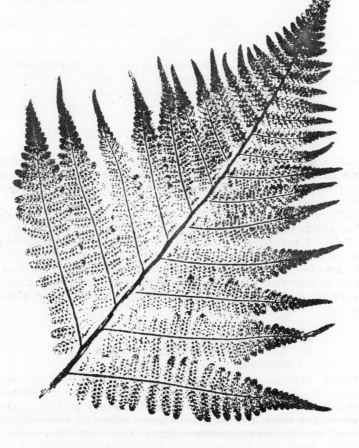

FOOD Quileute, Makah, Quinault, Squamish, Klallam, Samish, Sechelt, and Mainland Comox steamed rhizomes overnight and peeled them before eating, sometimes with oil or dried salmon eggs. Squamish, Klallam and Salish rubbed the fuzz off fiddleheads and ate them raw or steamed. The mature plant above ground was never used as food. Mohney says rhizomes, though edible, are not practical, being small, hard to clean, and not tasty.

MEDICINE Cowlitz treated body pains with a rhizome decoction. Makah pounded stems of four ferns to make a tea to ease labor. Old herbalists used powdered rhizomes to heal ulcers. Traditional and herbal medicine formerly used Lady Fern rhizomes as a vermifuge.

OTHER Shuswap made a black dye from Lady Fern. Kwakiutl combined fronds with Indian Paint Fungus to make a general purpose red paint. Lady Fern can be aggressive, but can also be useful in a home garden; it is happy in the open if well watered.

LADY'S SLIPPER, FAIRY, or ANGEL SLIPPER
Calypso bulbosa   Orchidaceae

NAMES *Calypso*, from the Greek sea nymph *Kalypso*, means covered or hidden from view, referring to the plant habit of hiding among forest mosses; *bulbosa* describes the corm or bulb. This genus has only one species.

FOOD Haida and Lower Lillooet ate corms, raw, roasted, or boiled; they had access to old-growth forests, a habitat suited to its growth.

OTHER This Orchid of shady woodlands, blooming from April from mosses or in thick deposits of Douglas Fir or Hemlock needles, grows only in association with certain fungi, is pollinated by only a few insects. It will not long survive picking or transplanting, and experts have tried it; if you are fortunate enough to find this dainty Orchid, enjoy it where it grows.

LAMB'S QUARTERS, PIGWEED
Chenopodium album   Chenopodiaceae

NAMES Pigweed says pigs enjoy eating this ubiquitous annual garden weed, native of Eurasia, which blooms from June to September. The genus name, from the Greek *chen*, goose, and *pous*, foot, refers to leaf shape. The white, *album*, of the species name, for the scaly white underleaf, is part of another common name, White Goosefoot.

FOOD Peoples of many times and cultures have used leaves, tops, and seeds of Pigweed as emergency or basic foods, and wildfood authors still recommend this; related species are similarly used. Excavations of early lake villages in Switzerland confirm that it was eaten in antiquity. The greens, high in protein, are a good source of vitamin C, an even better source of A; it is a good source of iron and potassium and extremely rich in calcium. Many people prefer this to spinach, young leaves as salad, older leaves cooked. Flower buds and flowers can also be cooked as potherbs. Seeds can be ground as flour for use in bread or cooked as mush. Seeds can be eaten without grinding and are often part of *pinole*, a Spanish term for a meal made from a mixture of seeds of small plants.

Based on research in the latter 19th century, Sturtevant reported Indians of New Mexico, Arizona, California, and Utah ate Pigweed raw or as a potherb, and used seeds ground to flour. Food use by Northwest Washington and British Columbia Coastal groups was not reported, but British Columbia Interior Lillooet, Okanagan, Shuswap, Flathead of Montana, Cheyenne and Indians farther south ate seeds pounded into flour and made into bread.

MEDICINE Stuhr says Indians used the leaves of Lamb's Quarters to relieve stomach ailments. Pollen of some *Chenopodium* species may cause hay fever and other allergic responses.

OTHER A moss green dye has been made from Pigweed with alum, copper, or iron.

LAUREL CHERRY, CHERRY LAUREL
Prunus laurocerasis   Rosaceae

NAMES The common name translates the species, *laurus*, Laurel, and *cerasus*, Cherry. *Prunus* is the ancient Latin name for plum. This favorite evergreen hedge shrub, with whitish flowers at stem tops in April, sometimes escapes to Northwest woodlands. It was brought from Asia Minor to European gardens by 1580.

FOOD A volatile oil obtained from the leaves was formerly substituted as a less expensive food flavoring than Oil of Bitter Almonds. Such use is dangerous as leaves, twigs, and fruit pits contain cyanogenic glycosides. Acid content varies with growing conditions, but poisonings have resulted from food use.

OTHER Shoots of *P. laurocerasis* produce a reddish brown dye. We call this shrub Laurel, but the true Laurel is *Laurus nobilis*, or Sweet Bay, in the family Lauraceae. Leaves of Sweet Bay are the Bay leaves used in cooking.

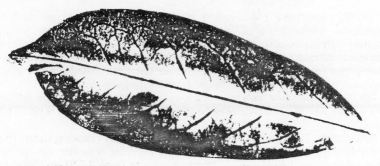

Laurel Cherry
Prunus laurocerasis

LEMON BALM
Melissa officinalis   Labiatae

NAMES The crushed leaf has a tangy lemon scent; balm is an abbreviation of balsam, chief of the sweet-smelling oils. *Melissa,* Greek for honeybee, identifies this perennial as a honey source; *officinalis* says Linnaeus, who named it, knew of medicinal use. Lemon Balm, a native of Mediterranean regions, had reached Britain by 1573. This easily-cultivated herb blooms from June. It occasionally escapes to wild areas.

FOOD Dried or fresh leaves make a beverage tea. Leaves can be used in salads, soups, and sauces, and are often included in potpourri.

MEDICINE Pliny, Dioscorides, and Gerard ascribed virtues to this plant. Herbalists have used Lemon Balm as diaphoretic, febrifuge, carminative and emmenagogue. A leaf infusion has been used for colds; leaves, preferably fresh, were formerly applied externally to bites and stings. The plant, in the U. S. Pharmacopoeia in 1890, contains some tannic acid and small quantities of a highly-flavored essential oil.

OTHER Many insects, including mosquitoes, are repelled by the scent of Lemon Balm. It serves as an insecticide in gardens, as do several other culinary herbs.

Prickly Lettuce, Compass Plant
Lactuca serriola

PRICKLY LETTUCE, COMPASS PLANT
Lactuca serriola (L. scariola) . Compositae

NAMES Compass Plant refers to leaves of this Eurasian native, which grow vertically against the stem in a generally north-south direction. Prickly is also descriptive. *Lactuca* comes from *lac,* ancient Latin name of milk, as plant stems have milky juice. Prickly Lettuce has bluish-green foliage and yellow flowers. Air currents distribute seeds of these annual or biennial plants to fields and woods.

FOOD Wildfood editors say nine western *Lactuca* species are edible when young, raw in salads or cooked as potherbs; older plants tend to be bitter.

MEDICINE Dried juice from the roots is used as chewing gum. The dried milky juice of plants in the genus *Lactuca* was the source of Lactucarium, listed in the U. S. Pharmacopoeia from 1820 to 1926. *L. virosa* was official, but *L. serriola* was also used. Garden Lettuce, *L. sativa,* is said to have lesser quantities of similar properties. Herbal medicine prescribed Lactucarium as diaphoretic, diuretic, narcotic and sedative. This was formerly an ingredient in proprietary preparations used to treat coughs and quiet nervous irritations; these became less popular after morphine was removed from the contents. Lactucarium was sometimes mixed with brandy as a medicine; one reference says any benefits from Lactucarium were purely psychical. Prickly Lettuce, used as a source of sedatives since antiquity, is not presently regarded as effective for this purpose. It may have been thought to have sedative power because taste and odor of its bitter juice are similar to that of opium, for which it has been used as an adulterant.

OTHER   Young Prickly Lettuce consumed in large quantities is toxic to cattle. A 1786 dye list said Prickly Lettuce provided a dye the color of wool produced in Vigogna, Spain.

WILD LETTUCE
Lactuca biennis   Compositae

NAMES   Lettuce may be a variant of *Latues,* the Old French word for lettuce, from the Latin *Lactuca. Biennis* says Wild Lettuce is a biennial, but its bluish white or pale yellow blooms sometimes appear during its first year.

FOOD   Bitterness of Wild Lettuce reduces its appeal for salads or potherbs, as even the young plants have a bitter aftertaste.

LICORICE FERN
Polypodium glycyrrhiza (P. vulgare)   Polypodiaceae

NAMES   Licorice Fern roots taste somewhat like licorice; it is not as sweet as candy, but has been used for many years to make cough syrups and medications more palatable. *Poly,* Greek for many, plus *pous,* foot, refers to knoblike branches on the rhizome. *Glycyrrhiza,* from the Greek *glykys,* sweet, and *rhiz,* root, describes root flavor.

Linnaeus identified the European Licorice Fern as *Polypodium vulgare;* this name was formerly used also for North American plants. As North American species are now believed to differ from the European, the name has been changed to *P. glycyrrhiza.*

The entire width of Licorice Fern leaflets or pinnules is attached to the stem; Sword Fern leaflets join the main stem at the center vein only. Fronds of Licorice Fern grow singly or in groups in moss on rocks, logs, or tree trunks, especially those of Bigleaf Maple but also on Alder and Oregon White Oak. Occasionally Licorice Fern grows on an Elderberry old enough to have a thick layer of moss. This almost-evergreen fern extracts nutrients from the air, taking only housing from its host, and can be found all year long if water is available. Fronds, dependent upon moisture held by mosses in which the roots grow, die back after a long dry period and reappear after substantial rains.

Licorice Fern
Polypodium glycyrrhiza
small frond; tip of larger frond showing spore pattern

110

FOOD South Kwakiutl ate small quantities of the rhizomes, steamed, in times of hunger. Saanich used them to sweeten food; some other British Columbia groups chewed them for flavoring. Calapooya ate raw rhizomes after removing the outer bark. Early settlers used the root to flavor tobacco; it has been an ingredient of plug tobacco. In the search for a commercially feasible low-calorie replacement for sweeteners, a substance 30 to 300 times as sweet as sugar was identified in rhizomes of *P. vulgare*. The yield is low and this has not been tested for toxicity.

MEDICINE Western Oregon Indians pounded Licorice Fern rhizomes, steeped them in hot water and used the liquid, strained and sweetened, to treat dysentery. Several British Columbia coastal groups treated colds and sore throats with the rhizome. Makah chewed peeled roasted rhizomes and swallowed the juice for coughs. Quinault, Klallam, and Green River chewed baked roots or used them raw as cough medicine. Early settlers used rhizomes as asthma and catarrh remedies.

Indian, traditional, and herbal medicine all record use of Licorice Fern rhizomes for cough medicines. The ancients valued Licorice Fern as a purgative. Herbal and traditional medicine have used root preparations as diuretic, demulcent, anthelmintic, and purgative. It was listed in the U. S. Pharmacopoeia for many years, sometimes for flavoring other medications.

OTHER Licorice Fern is nice in a wild garden where mossy banks and trees provide a suitable habitat; it volunteers in hanging baskets from spores in the moss. The Licorice of commerce, *Glycyrrhiza glabra*, is an entirely different plant, belonging to the Pea family.

LUPINE

Lupinus species   Leguminosae

NAMES *Lupus*, Latin for wolf, reflects an early belief that these plants robbed soil of fertility. Pea family members are now known for the ability to fix nitrogen, helped by bacteria associated with their roots. Some Lupines will grow on poor soil, and all enrich the soil; some are annuals, some perennials. Lupines have been cultivated since the days of the ancient Egyptians as forage and as green manure.

FOOD Early Romans cultivated a white Lupine, which Virgil called "sad Lupine" as the poor ate the seeds, first boiling them to remove the bitter taste. Current authors say eating quantities of Lupine seeds is dangerous because they contain poisonous alkaloids. About four dozen species are native to Oregon; though not all are known to be toxic, it seems prudent to avoid eating them.

Some Northwest Washington, Oregon, and British Columbia Indians peeled, roasted and ate roots of *L. littoralis*, Shore Lupine; these were sometimes eaten with grease. Reported food use is limited to this one species, although some animals apparently eat Lupine seeds of many species without problems. Some California Indians ate spring leaves and flowers of a Lupine and used seeds for a medicinal tea.

MEDICINE   Parkinson said Lupines had virtues. Herbalists formerly used seeds as anthelmintic, diuretic, emmenagogue, as a demulcent and as an emollient poultice to treat local inflammations.

OTHER   Lupines were used to anchor sand dunes when grounds were prepared for Golden Gate Park in San Francisco. Many cultivated Lupines are available, probably more desirable for home gardens than the wild species.

MADRONA, MADRONE
Arbutus menziesii   Ericaceae

NAMES   Early Spanish-Californians called this Madrona because it resembled a Spanish tree called *Madrono. Arbute* was the Latin name of a tree known in early Roman times. Menzies shares his name with other Northwest native plants. Madrona requires sharp drainage and often grows in rather poor soil.

FOOD   Takelma of Southwest Oregon and some California groups reportedly ate Madrona fruits. Wildfood editors recommend these raw, cooked, or dried, as they keep well; they are said to be edible but seedy, and bland in flavor. Birds like the fruit.

MATERIALS   Saanich and Straits Salish made spoons and gambling sticks from young Madrona branches. Recently Sechelt used wood for sterns and keels of small boats, as Madrona is durable under water. Saanich and Cowichan boiled bark for tanning paddles and fish hooks. Wood is considered good for furniture, although it is not widely available at commercial outlets.

MEDICINE   Skokomish drank a leaf decoction for colds and sore throat; sometimes roots of Oregon Grape or Licorice Fern were added. Stuhr said Indians used an astringent infusion of bark, root, and leaves of Madrona to treat colds.

OTHER   Smooth reddish-brown bark and broad evergreen leaves make this desirable for well-drained garden sites. Leaves last about a year and a half. Trees as old as 250 years have been found. Those from 30 to 50 feet tall and 15 inches through are probably about 100 years old.

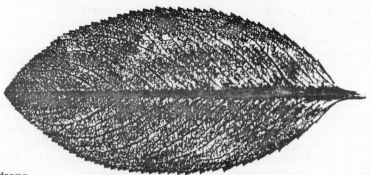

Madrona
Arbutus menziesii

MAIDENHAIR or FIVE FINGER FERN

Adiantum pedatum   Polypodiaceae

NAMES   Maidenhair may refer to the black stems, which some think resemble the shiny hair of Indian maidens, to the fine black fibrous roots, or to use of Maidenhair products for hair care. Five Finger Fern describes the palmate leaf pattern of this species. *A* meaning not, and *diane,* to wet, say the foliage sheds rain; *pedatum,* footlike, says leaf branches are shaped like the feet of birds.

MATERIALS   Makah, Quinault and Hoopa used the stems for contrast design in fine baskets. For drying berries, many groups covered Cedar-bark mats with Maidenhair Fern, as the thin leaves were easy to winnow out and remnants did not appreciably affect taste of dried berries.

Maidenhair Fern
Adiantum pedatum

MEDICINE Makah chewed leaves to treat sore chests, stomach trouble, and internal bleeding. Herbal and traditional medicine formerly used the whole plant, preferably fresh, for coughs, asthma, treatment of pulmonary ailments, and as an expectorant. Preparations from the European Maidenhair were used as an emmenagogue. Reported medicinal use of Maidenhair leaves in Europe, China, Ethiopia, and Central America shows wide distribution of *Adiantum* species.

Makah, Lummi, and Skokomish soaked leaves in water used for washing hair. Quinault rubbed ashes of burned leaves on hair. Gerard recommended Maidenhair for hair care, and recent herbalists have suggested a final rinse with a strained, cooled plant tea to add body and sheen to hair.

OTHER Maidenhair Fern grows well and makes an attractive addition to a shady moist garden. Dried fronds are used in floral arrangements.

BIGLEAF MAPLE
Acer macrophyllum   Aceraceae

NAMES Common and species names say the same thing in different languages: Greek, *macro*, large, and *phyllon*, leaf. *Ac* is Celtic for hard or sharp, so *Acer*, Latin name for Maple, was a good choice for a tree with hard wood valued for making weapons. This deciduous tree has the largest leaves of any native tree in our area, commonly six to ten inches long, with five lobes of about equal width. Wind pollination of the drooping yellowish-green blooms is aided by the absence of leaves, just unfolding in early April. Maple seeds, also called keys or samaras, grow in pairs, with the angle of attachment varying in different species. Bigleaf Maple grows mostly west of the Cascades; its wings are attached in a V shape at an angle of less than 90 degrees. Height may exceed 75 feet; trees 50 to 60 feet tall with trunks two to three feet in diameter are common; these may be from 50 to 85 years old. Bigleaf Maple trees are mature at 100, but may live to be 200 years old.

FOOD and MEDICINE Some Coastal British Columbia groups boiled and ate sprouted Maple seeds; most did not eat Maple cambium. Klallam used a bark infusion for tuberculosis. Birds, squirrels, and other small animals eat the seeds.

MATERIALS Haida, Tlingit, Tsimshian, Quinault carved bowls, dishes, platters, and spoons from Maple. Klallam, Snohomish and Skagit made canoe paddles; Swinomish and Lummi made cradle boards; Cowlitz used bark for ropes and tumplines, Skagit, Lummi and Snohomish used Maple leaves in cooking pits and to cover food. Squamish did not use Maple for cooking, but laid fish on Maple leaves for cleaning. Squamish, Swinomish, Chehalis, and Quinault smoked salmon with decayed Maple wood; some Coastal British Columbia groups laid drying berries on Bigleaf Maple leaves. Leaves were also used as wiping materials. Kwakiutl made baskets from the inner bark. Bigleaf Maple is good fuel as it burns hot and with a smokeless flame. Sunset magazine recently reported that Maple was the second most important Northwest hardwood; it is used for cabinets and furniture.

Bigleaf Maple
Acer macrophyllum

OTHER   Maple sugar is often made from the sap of Silver Maple, of eastern North America, *A. saccharinum*, which produces greater quantities of sap than other species. Sap flows best on sunny days after a period of freezing weather. Thirty to forty gallons of sap are boiled down for one gallon of syrup, so even with trees which produce efficiently, making syrup is hard work. It is theoretically possible to make sugar from any Maple species, but not too practical here.

MEADOWRUE, WESTERN THALICTRUM

Thalictrum occidentale   Ranunculaceae

NAMES   Linnaeus chose the Greek *thaliktron,* used by Dioscorides for some unknown plant, for this genus; *thallo* in Greek means to become green, *occidentale,* western. This lacy native blooms in April in moist areas. Its inconspicuous flowers, male on one plant, female on another, have neither petals nor perfume, suggesting wind pollination.

MEDICINE   Eastern Oregon Washoe and some Nevada groups treated colds with a root tea. Washoe, Paiute, Shoshone, Warm Springs and some California Indians made a shampoo from dried roots. Warm Springs, Shoshone, Blackfoot, and Gros Ventre used seeds or the dried whole plant for perfume. Blackfoot chewed dried seeds and rubbed them on skins and body as a perfume.

OTHER   The pretty foliage mades Meadowrue desirable for native gardens.

MINER'S LETTUCE

Montia perfoliata   Portulacaceae

NAMES   This genus, found chiefly in North America, honors an early Italian botanist, Guiseppe Monti, 1682-1760. The stem seems to grow through two leaves which are joined just below the flower, hence *perfoliata,* through the leaf. This delicate herb blooms in March, then dies back. Wildflower books consistently suggest Miner's Lettuce as the common name of this herb.

FOOD   Many western Indians ate roots, stems, and leaves, raw or cooked. Wildfood editors recommend the same uses. Antiscorbutic properties of Miner's Lettuce made it valuable to early miners, who had limited supplies of fresh food. This native American plant was introduced and cultivated in Europe in the early 19th century; it was called Winter Purslane and eaten raw or cooked.

MEDICINE   Snohomish, Quileute, Skykomish, and Cowlitz rubbed the stem between their palms, and rubbed the plant in water for a tonic to make hair glossy. Quileute treated dandruff with the same solution.

OTHER   Many name changes for this genus reflect differing opinions of botanists. Hitchcock says present practice classifies plants with rootstocks as *Montia,* those with corms as *Claytonia.* In our area, *Claytonia* species usually grow at higher elevations.

MOCK ORANGE or SYRINGA

Philadelphus lewisii   Hydrangaceae

NAMES   Mock Orange blossoms resemble those of the Orange; Syringa is another name for this plant family. The genus honors Philadelphus, the Egyptian pharoah Ptolemy II, who lived from 308 to 246 B. C.; the species, Meriwether Lewis.

MATERIALS   Many Indians made arrowshafts of Mock Orange, both for hunting and war, and used older wood for bows. Cowlitz and Snohomish used leaves for a soap substitute, which Snohomish also rubbed on sores as a medicine.

OTHER   This fragrant shrub, the Idaho state flower, is recommended for use in wild gardens where it can be enjoyed without picking, as the showy white flowers wilt rapidly when cut. It blooms in sunny places in open woods or rocky slopes, and is attractive to butterflies. Mock Orange has long been valued in England as an ornamental; David Douglas introduced it there in 1825.

MONKEY FLOWER
Mimulus species   Scrophulariaceae

NAMES   An imaginative individual can see a monkey face in this flower, providing both common name and a tie to the genus, the diminutive of the Latin *mimus*, a mimic. *M. guttatus*, a native of western North America, has showy yellow blooms up to 1 1/2 inches long, which brighten moist streambanks or wet places from May. *Guttatus*, spotted or speckled, refers to spots on the lower lip of the flower of this annual or perennial. This species varies greatly under different growing conditions; it may grow from 2 to 24 or more inches tall. Its leaves are palmately veined.

M. moschatus, also yellow, moisture-loving, and blooming in May or June, may have a more or less musky fragrance, translating the Latin *moschatus*. The sticky-feeling leaves of the perennial *M. moschatus* have pinnate veins; the yellow flowers have unmarked lips. Both grow in the Portland area. The perennial *M. dentatus*, toothed, is only found near the coast; its veins are also pinnate, the major lateral veins arising below the middle of the leaf.

FOOD   Kirk recommends *M. guttatus* for salad and calls it Wild Lettuce. Perhaps stickiness of other species makes them undesirable.

MEDICINE   In areas where Indians had horses, raw leaves and stems were used to treat rope burns.

OTHER   Pink, purple, or yellow flowers of other *Mimulus* species grow in moist, often alpine, areas. *M. guttatus* is nice for wet ground in a sunny wild garden. This rather large genus is best developed in Western North America; Hitchcock lists nineteen species in the Pacific Northwest.

Monkey Flower
Mimulus guttatus
palmately veined leaf

MORNING GLORY, BINDWEED .
Convulvulus arvensis   Convulvulaceae

NAMES Morning Glory says this flower opens early; it closes in the dark. Bindweed may reflect use of its network of tough roots in India to reduce erosion, as they kept the sea from washing away the sand. In an equally logical but less kindly interpretation, Bindweed identifies its habit of binding tendrils around plants as it climbs up for light and air. *Convulvere* means to twine, and this vine certainly does. Latin *arvensis*, of the fields, identifies the plant's preference for a somewhat open habitat; it is one of the most serious field weeds.

The widespread food-storing rhizomes of this Old World perennial resist removal, and even a small remnant sends up a sprout. The five stiff tracks of the bell-shaped blossom mark out a path for bees seeking nectar.

OTHER Morning Glory shoots can be used to dye wool a musk color. This plant may poison livestock where potassium nitrate is in the soil, as the seeds absorb nitrates in toxic quantities. Seeds of horticultural Morning Glory are reported to have similar toxins.

MOUNTAIN ASH, ROWAN
Sorbus (Pyrus) aucuparia; S. sitchensis   Rosaceae

NAMES *Sorbus* species are not truly Ashes, which belong to the Olive family. Perhaps the divided leaves suggested the name, for Ash also has compound leaves. *Sorbus* is another early Roman name with no known relationship to its present use. Birds have spread *S. aucuparia,* native of Sweden, to many wild areas. This ornamental, which may grow to be a tree, usually has 13 or more leaflets, at least in part. The native Mountain Ash, *S. sitchensis,* was named for Sitka, Alaska, source of the original description. It is more common at higher elevations, and is usually a shrub; the leaves have 7 to 11 leaflets.

FOOD Mohney says Mountain Ash was a reasonably popular food of early Indians, though this was not mentioned in other sources. Wildfood editors recommend Mountain Ash berries of either species, especially after frost, raw, cooked, or dried, for jams, jellies, in pie filling, or in sweet wine. Fruits of *S. aucuparia* have been used in Scotland for marmalade, and have been roasted for a coffee substitute. The sweetening agent sorbitol occurs naturally in small quantities in ripe Mountain Ash berries, apples, and other fruits. In Northern Europe Mountain Ash berries have been dried for flour, and fermented to make a strong drink; the Welsh formerly used these to make ale. The sour, bitter fruit is a favorite of birds; bears eat them, too. Deer and elk browse on young twigs.

MEDICINE Bella Coola rubbed berries in the scalp to treat lice and dandruff. Herbalists recognized all parts of the European Mountain Ash as astringent. Bark and fruit, fresh or dried, were valued for medicine. Ripe berries are antiscorbutic and provide a gargle for sore throat; bark preparations were used for amenorrhea and leucorrhoea.

OTHER   British Columbia Carrier used Mountain Ash to make side sticks for snowshoes. Depending on mordants, fruits have provided dyes of greenish-yellow, grayish-brown, and red. All parts have been used to provide a black dye, and bark has been used in tanning. Native Mountain Ash has been recommended as a garden shrub; at lower altitudes it bears fewer blooms than in its mountain home—fewer berries to clean up! The European tree is much used in horticultural plantings.

MULLEIN
Verbascum thapsus   Scrophulariaceae

NAMES   At one time Mullein was considered a remedy for leprosy, Latin *malandrium,* the Anglo-Saxon *moleyn. Verbascum,* from *barbascum,* Latin for bearded, says the plant has shaggy foliage. An ancient town named Thapsus—either in North Africa or in Sicily— provided the species name. The flowers, quite different from those of its relatives Foxglove and Snapdragon, appear a few at a time from June. Leaf arrangement of this plant of dry open areas conserves rain water by directing it to the next lower leaf, then to the root system. Wooly hairs on leaves and stems help maintain moisture and frustrate creeping insects seeking nectar or pollen; these scratchy hairs also discourage browsing animals.

MEDICINE   Gerard, Culpeper and Parkinson regarded Mullein more highly than later practitioners. Herbalists formerly used leaves and flowers, or roots, preferably from wild plants, as demulcent, emollient, antispasmodic, diuretic, and astringent. Leaves were boiled in milk for a diarrhea treatment. Coughs and chest infections were sometimes treated by smoking dried leaves. Mullein tea, an old remedy for colds and coughs, was strained to avoid irritating throat tissues with the small hairs. The British Pharmacopoeia and the U. S. National Formulary formerly included flowers in chest remedies, and dried leaves for a demulcent. Early Californians applied leaves to treat sprains and lung diseases.

OTHER   Early Spaniards dried and smoked the leaves, a use adopted by Hopis in the American Southwest. Dried down on stems has been used for tinder. In Imperial Rome, bloom spikes of this Eurasian biennial weed were dipped in tallow for use for torches. Yellow dyes have been made from dried leaves and roots. Early Roman ladies used roots for a yellow hair dye; soap from ashes was supposed to restore gray hair to its original color. The velvety gray-green leaves provide interesting color contrast in flower arrangements.

YELLOW or FIELD MUSTARD
Brassica campestris   Cruciferae

NAMES   Ancient Romans ate Mustard seeds pounded and steeped in new wine, explaining the common name: *mustum,* newly fermented grape juice, and *ardens,* burning. *Brassica* is the Latin word for Cabbage; important foods supplied by this genus include Cabbage,

Broccoli, Turnips, and Cauliflower. *Campestris* translates as of the fields, a habitat brightened by Yellow Mustard blooms in March and April. Flowers of this European annual weed, escaped from farmers, attract bees.

FOOD Wildfood editors recommend young greens of various Mustards as a potherb, and seeds in moderation for seasoning. Oil in the seeds causes irritation for some people, and seeds in quantity are poisonous to stock. Mustard browsed by cattle causes an undesirable flavor in milk. Mustards have been used as forage for sheep and as green manure; seeds planted after harvesting an early crop grow in the same season, protect soil from wind damage and provide a helpful supplement when plowed under.

MEDICINE Hippocrates recommended Mustards, seeds for medicine and as a condiment, greens for a cooked vegetable. Yellow Mustard was not recommended. Seeds of Black Mustard, *B. nigra,* and White Mustard, *B. alba,* were formerly included in the U. S. Pharmacopoeia as irritant, stimulant, diuretic, and emetic. A principal use of Mustard by both traditional and herbal medicine was as a counterirritant poultice—the so-called Mustard plaster. In England, seeds of *B. nigra* have been chewed for toothache relief.

NEMOPHILA, GROVE LOVER
Nemophila parviflora   Hydrophyllaceae

NAMES The Greek *nemos* means grove, and *philein* means to love, both genus and common names identifying habitat of this native annual. The small flowers, *parviflora,* appear in May in the shady places it prefers. The inconspicuous whitish flowers are less interesting than the vine-like leaves of this woodland ground cover, which gets rather straggly for use in home gardens.

Nemophila, Grove Lover
Nemophila parviflora

PACIFIC NINEBARK

Physocarpus capitatus   Rosaceae

NAMES   Bark of this shrub peels off in nine or more layers; some say there are as many as twenty. The Greek *physa*, bellows or bladder, and *karpos*, fruit, describe the inflated seed capsules. *Capitatum*, Latin for head, pictures the head-like white flower clusters, appearing in late May in woodland edges, almost entirely west of the Cascades.

MEDICINE   Bella Coola and Green River made an emetic tea from inner bark or peeled young shoots.

OTHER   This is an interesting native shrub for wild gardens.

Ninebark
Physocarpus capitatus

NIPPLEWORT

Lapsana communis   Compositae

NAMES   Dioscorides used *Lapsane* for some Mustard family member; *communis* means common. Nipplewort blooms in June and July in gardens, fields, roadsides and disturbed sites. Like many other Sunflower family members, this yellow-flowered Eurasian straggler, common west of the Cascades, is not particularly attractive.

FOOD   Spring greens of this annual weed are used before flowering as a potherb or for salads.

OREGON WHITE OAK, GARRY OAK
Quercus garryana   Fagaceae

NAMES   Oaks were considered a fine, (Celtic *quer*), tree, (*cuez*), because the sacred Mistletoe grew on some species. Oak was a dominant tree in the later Mesozoic period. Some relict species date back to the warm interglacial period thousands of years ago. *Garryana* honors Nicholas Garry, secretary at the Hudson's Bay Company and a helper of David Douglas, who named the species. Oregon White Oaks grow mostly west of the Cascades in Washington and Oregon, seldom west of the Coast range; many grow in the Willamette valley. Leaves of Garry Oak have rounded lobes.

FOOD   All Oaks have edible acorns, some more palatable than others. White Oak acorns, which ripen in one growing season, are sweeter and less bitter tasting than those of Black Oaks, which need two seasons. Nisqually, Chehalis, Cowlitz, Klallam, Squaxin, and some British Columbia Salish groups ate acorns in small quantities. Because of tannic acid in acorns, it is important to use a leaching process before eating them in quantity. Where Oaks were common, as in much of California, acorns were a major food source. Northwest Indians, with fewer Oaks available, developed fewer leaching techniques. Calapooya ground acorns and left them in a creek long enough to wash out the tannin. Some California Indians buried acorns in swampy areas over winter, dug and ate them in spring. Another process used wood ashes as a kind of lye. Takelma of Southwest Oregon ground acorns to meal, soaked them, then cooked them in hazeltwig baskets to make mush. Tillamook also ground and used them for porridge.

Wildfood sources recommend acorns; small amounts can be eaten raw, but are better roasted, also safer. Current editors suggest percolating ground acorns two or three times with hot water. Leaching removes little of the essential food elements. Acorn meal is said to contain about 25 per cent fat, approximately 60 per cent carbohydrates, and 4.5 to 5.5 per cent protein. Acorns are a food source for many animals, a favorite of bandtailed pigeons. Oak leaves, also high in protein, are browsed by mule deer and other wild animals; they are toxic to animals if eaten exclusively.

MATERIALS   Cowlitz used the strong wood for digging sticks and combs. An excellent fuel, Oak burns clean and makes a lasting fire. Oaks are not sufficiently available in our area to be valuable for timber use; also, mature Oaks of other species tend to have longer trunks.

MEDICINE   Cowlitz treated tuberculosis with a boiled bark solution. Herbalists used the astringent Oak bark to treat diarrhea, and as a quinine substitute. Galen recommended applying bruised leaves to heal wounds. Oak galls, formed by larvae of tiny wasps, are used externally as an astringent; they contain proportionally more tannin than Oak bark. Some authors recommend that outdoor enthusiasts know the location of Oaks as a first aid source: a bark solution can be used to treat sore throats, burns and inflammations; chewed leaves can be applied to lessen pain of insect bites. Traditional medicine has recognized dried inner bark of an eastern White Oak, *Q. alba*, in the U. S. Pharmacopoeia as astringent, tonic, and hemostatic; it was mostly used externally. Pollen of some *Quercus* species causes hay fever.

OTHER Much of our native flora is thought to be derived largely from plants of northern origin; Oaks are one of the few believed to have originated in the warm climate of areas farther south. Black and White of the common names roughly describe color of bark of typical mature Oak trees; bark of Black Oaks is much darker than that of White Oaks. The Oregon White Oak is the only species relatively common here, as most prefer warmer locations. These hardy trees, native to drier areas of Western Washington, Oregon, and California, grow rather slowly to 40 to 90 feet, and may live from 100 to 500 years. Oaks are recommended for parks and for well-drained home gardens with spacious lawns.

Oak bark, mordanted with iron sulfate, makes a black dye; mordanted with alum, the dye is light brown or tan. Eight quarts of bark are needed to dye a pound of wool.

OCEAN SPRAY, ARROWWOOD
Holodiscus discolor   Rosaceae

NAMES Ocean Spray pictures the white bloom spire, appearing about June; Arrowwood reports a use for the straight stems. *Holo,* Greek for whole, and *diskos*, disk, describe the base of the flower parts; *discolor,* of two colors. Botanists call the bloom persistent, for brown, dried-up sprays hang on after color fades.

FOOD Quantities of the dry, hard, one-seeded fruits of this Northwest native, most common west of the Cascades, were eaten raw, cooked, or dried, by western Indians. The tiny seeds are hard to remove, though Mohney reports it is possible to make jam or jelly with the fruits, and suggests them as a field food, eaten without removing the seeds. Ocean Spray is browsed by deer and elk.

MATERIALS The name Ironwood reflects stem strength which made it desirable for arrow shafts, needles, sticks for digging clams and roots, prongs of flounder and duck spears, and other tools. It was often fire hardened. It is good for roasting tongs because it doesn't burn well. Pioneers used Ocean Spray pegs instead of nails.

MEDICINE Makah peeled Ocean Spray bark to make a tonic tea; Lummi prepared an eyewash from inner bark, a diarrhea treatment from blossoms, and put leaves on sore lips and feet; Squaxin used seeds with Wild Cherry to prepare a blood purifier.

OTHER Early specimens of *H. discolor* were collected by Meriwether Lewis, "on the banks of the Kooskoosky" (Idaho), and Menzies, "Northwest coast of America." This shrub is good for wild gardens and for reclaiming open or disturbed land.

Ocean Spray
Holodiscus discolor
end of branch

Tall Oregon Grape
Berberis aquifolium
pinnately veined leaflet

Low Oregon Grape
Berberis nervosa
two leaflets; palmately veined

124

LOW or LONG-LEAVED OREGON GRAPE
Berberis (Mahonia) nervosa   Berberidaceae

NAMES   Fruit and growth habit gave the common names. *Berberis* is from the Arabic name for the fruit, *berberys*. *Mahonia* honors Bernard M'Mahon, an early American botanist. See p. 2 for comments on genus names. *Nervosa* calls attention to the palmately-veined leaflets, variously described as having three to eight nerved veins, three central veins, or three to five veins arising from the leaf base. Another common name is Dull Oregon Grape; Hitchcock says leaves of this species are "somewhat glossy." For Tall Oregon Grape, he says "glossy."

Both Low and Tall Oregon Grape have similar uses; the species were not always separated by those reporting. Low Oregon Grape grows in shady open woods west of the Cascades; its compound leaves have 9 to 19 leaflets.

OTHER   Low Oregon Grape spreads by rhizomes, helpful for plants of shady woodlands which may be less likely to set seed. Some of its evergreen leaves turn scarlet in winter. This species is a good ornamental for shady or wooded areas.

TALL OREGON GRAPE
Berberis (Mahonia) aquifolium   Berberidaceae

NAMES   The common name refers to height of this handsome shrub which may reach six feet or more. *Aquifolium,* from the Latin *aqua,* water, and *folium,* leaf, refers to the leaves, so shiny they may appear wet. Both top and bottom sides of leaflets are shiny, giving rise to another name, Shining Oregon Grape. The bright yellow flowers unfold from March. Tall Oregon Grape, the Oregon state flower, has 5 to 9 or 11 pinnately-veined leaflets. Toward the end of a two- to three-year life span, the evergreen leaves may turn to bronze, vermillion, or orange.

FOOD   Cowlitz, Lummi, Makah, Upper Skagit, Lower Chinook, Samish, Swinomish, Snohomish, Squaxin, Western Oregon and British Columbia groups formerly ate large quantities of Oregon Grape berries, raw or cooked, in pemmican, to flavor soups, dried for winter, sometimes mixed with other berries for drying. Many now use Oregon Grape for jams and jellies. Willamette Valley Indians did not dry these berries, but ate them fresh. Wildfood editors suggest very young leaves or small quantities of flowers for salad, as well as berries for jelly, sauce, or wine, perhaps combined with Salal or Apple. Berries can be used for beverage teas; campers enjoy lemonade flavored with Oregon Grape juice. Birds, squirrels and chipmunks, and other small animals eat the berries.

MEDICINE   Cowlitz washed sores with a bark decoction. Quinault treated coughs and stomach disorders with a root decoction. Swinomish and Samish made a root tea for a general tonic. Squaxin gargled a root tea for sore throat and drank a similar tea for a spring blood purifier. Some Vancouver Island Salish prepared a root extract as a lotion for washing hands. California Indians used root or bark solutions for stomach trouble, as cough medicine, tonic, and to treat ulcers. Liquid from chewed roots was put on injuries or wounds, and a root decoction was used to wash cuts and bruises.

Herbal and traditional medicine have considered dried rhizome and roots of Oregon Grape alterative, diuretic, antiperiodic, bitter tonic, and laxative. Fresh berries and flowers, or dried roots, have provided a tea for a laxative or treating fevers. *B. aquifolium* was listed in the U. S. Pharmacopoeia from 1905 to 1916 and has been included in the National Formulary, most used as a bitter stomachic. Dried stems and leaves contain the drug berberis and the alkaloid berberine. Commercial supplies in the United States were obtained mostly from Oregon and Washington, as the Pacific Northwest natives *B. nervosa* and *B. aquifolium* were the main sources.

Creeping Berberis, *B. repens*, grows east of the Cascades; it has 5 to 7 or 9 leaflets; lower surfaces of its leaves are always dull; the upper may be glossy or dull. *Repens*, creeping, says the plant spreads, in this case by sending down roots along its stem.

OTHER   Oregon Grape is a well-known dye plant. Indians boiled chopped dried roots to make a yellow dye which was an important staple, used for basketry materials such as Bear Grass and porcupine quills, for fabric and buckskins. Makah, Chehalis, Cowlitz, Snohomish, Skagit, Klallam, Upper Skagit, Cowichan, Straits Salish, Nootka, and Okanagan used such a dye. A blue dye for wool has been made with berries, using alum or vinegar as a mordant.

Oregon Grapes, attractive at all seasons, are widely used in wild and cultivated gardens, and provide shelter and food for small animals. Some call them Holly Grape. Next to Salal, Oregon Grape is the most common evergreen shrub west of the Cascades. Tall Oregon Grape, often used in highway landscaping, grows best where there is sun. David Douglas introduced Oregon Grape to England in 1823.

OXALIS, REDWOOD SORREL; WESTERN YELLOW OXALIS
Oxalis oregana; O suksdorfii   Oxalidaceae

NAMES   The Greek *oxys*, sour, refers to oxalic acid in juice of these native Oregon perennials. Flowers, white or pinkish for *O. oregana*, yellow for *O. suksdorfii*, appear from April. W. N. Suksdorf was a botanist. Clover-like leaves fold in bud in an interesting shape; flowers open to light and sun, close at night and on dark days.

FOOD   Cowlitz ate leaves fresh or cooked; Quinault ate them cooked. Wildfood editors recommend leaves of *Oxalis* species for nibbling, for limited use in salads, combined with other greens as a potherb, and in "green rhubarb" pie. As it is high in oxalic acid content, Oxalis should be eaten sparingly. Its oxalates make it dangerous to livestock which eat large quantities.

126

Western Yellow Oxalis
Oxalis suksdorfii

MEDICINE Cowlitz sqeezed fresh juice of *O. oregana* plants into sore eyes; Quinault chewed roots and put the liquid into eyes. Quileute treated boils with wilted leaves. Herbalists have used plants of *Oxalis* species as acidulous, diuretic, antiscorbutic, and refrigerant.

OTHER Some members of this genus are pests, but Western Yellow Oxalis is nice for wild gardens; it is less aggressive than the white-flowered Oregon Oxalis. *O. stricta*, a troublesome garden and lawn weed west of the Cascades, has several yellow flowers much smaller than those of *O. suksdorfii;* it is thought to be native to North America.

OXEYE DAISY

Chrysanthemum leucanthemum   Compositae

NAMES The Greek *chrys,* gold, and *leuc,* white, combine with *anthemon,* flower, to make the scientific name gold flower, white flower—a confusing combination selected by Linnaeus.

FOOD Young leaves have been used for salad or potherbs, but are not currently recommended. Pretty flowers of this Eurasian perennial, escaped long ago from gardens, cover roadsides and fields from late May on into July. Grazing animals avoid this bitter plant if other food is available, and will crop grasses to the roots. In effect, this is protection which encourages spread of the Daisy.

MEDICINE Northwest Indians recognized Oxeye Daisy as non-native, but it was introduced early. Quileute boiled dried stems and flowers to make a wash for chapped hands. Culpeper regarded it as a wound herb of good repute for inward and outward wounds. Recent herbalists have used the whole herb and dried flowers as antispasmodic, diuretic, and tonic, and for a lotion for wounds, bruises, and ulcers.

European herbalists suggested fresh or dried leaves or flowers to drive away fleas. The insecticide Pyrethrum, obtained from dried flower heads of other Chrysanthemum species, has been included in the U. S. National Formulary. The Dispensatory says that *C. leucanthemum* has been the most common adulterant of Pyrethrum. This would indicate that its insect-repelling properties are less strong. Efficiency is determined by the number of insects destroyed by a given quantity of powdered dried flowers in a measured time period.

OTHER Luther Burbank used the Oxeye Daisy in hybridizing the Shasta Daisy, which has double flowers. Contact with pollen and flowers of Oxeye Daisy is suspected of causing hay fever and other allergic reactions.

127

Oxeye Daisy
Chrysanthemum leucanthemum
basal leaf

Pathfinder
Adenocaulon bicolor
appreciably reduced leaf

Perennial Pea
Lathyrus latifolius
tip of stem, showing tendrils, appreciably reduced

PATHFINDER, SILVERGREEN, TRAIL PLANT
Adenocaulon bicolor Compositae

NAMES Pathfinder, Trail Plant, and Silvergreen are descriptive. Walking through patches of this native perennial makes a trail, as bent stems expose white hairs on under sides of the large thin leaves. This is the *bicolor*, two colors, of the species and the Silvergreen of the common name. Greek *aden*, gland, and *caulon*, stem, say the stems have glandular hairs. Odor of the inconspicuous small flowers of Pathfinder attracts flies for pollination; its hooked seeds get around by hitchhiking.

MEDICINE Cowlitz and Squaxin applied bruised leaves to boils or sores.

OTHER Pathfinder has been suggested for wilder parts of shady woodland gardens.

WILD PEA, PERENNIAL PEA
Lathyrus latifolius Leguminosae

NAMES *Lathyrus* is an ancient Greek name for some Pea family member; *la* plus *thoursous*, something exciting, as it was thought to have medical qualities; *latifolius* says broad leaf. This European escape from cultivation is the most common weedy Sweet Pea. It is related to Vetches, *Vicia* species. Peas usually have larger leaflets and flowers. Both Vetches and Peas have a twining habit.

FOOD Beach Pea, *L. japonicus*, is the only *Lathyrus* species wildfood authors recommend for food; seeds of most *Lathyrus* species are toxic. Sweet Peas are also toxic, definitely not for eating. Edible peas are in the genus *Pisum*.

OTHER The hot pink flowers provide bright accents over a long blooming period. The extensive root system of Perennial Pea is valuable for neglected or disturbed sites; it helps prevent soil slippage or erosion on steep banks.

PEARLY EVERLASTING
Anaphalis (Gnaphalium) margaritacea Compositae

NAMES *Anaphalis*, an ancient Greek name for some everlasting flower, comes from the Greek *a*, not, plus *knaphalon*, wool or felt, a negative description of the dry, not wooly, flower heads. *Margaritacea* means pearly. White parts are modified leaves or bracts; the yellow center parts are the true flowers. A cottony substance on the wiry stems repels ants which try to climb to the plant's nectar.

MATERIALS Thompson stuffed pillows with dried flower heads. The dried plant has been used as smoking tobacco.

MEDICINE Quileute used Pearly Everlasting in steam baths to treat rheumatism. Herbal and traditional medicine formerly used flowers and stalks as astringent, pectoral, and as a poultice for sprains, bruises, and swellings.

Pearly Everlasting
Anaphalis margaritaceae

OTHER   This perennial everlasting, a North American native, is suggested for open sunny sites in wild gardens. Flower heads dry well, providing a common name and a desirable addition to winter bouquets.

PHANTOM ORCHID

Eburophyton austiniae (Cephalanthera austinae)   Orchidaceae

NAMES *Eburo,* Latin for woody, and *phyton,* Greek for plant, provided the currently-accepted genus name for this saprophytic Orchid. *Austiniae,* (the feminine form), honors American botanist R. M. Austin, 1823-1919. White leaves, stems, and flowers gave the Phantom of the common name. Hitchcock recommends *Cephalanthera* for a European genus which includes plants with green foliage.

Single bloom stems of this white Orchid, native to western North America, appear in coniferous woods in late June and early July.

PIGGY-BACK PLANT, YOUTH-ON-AGE

Tolmiea menziesii   Saxifragaceae

NAMES Piggy-Back, Youth-on-Age, and Thousand Mothers describe vegetative reproduction of this native perennial of moist woodlands, which develops small leaves at the base of large leaves, which later bend down and develop roots. Historians will recognize names of botanists with Northwest connections: Dr. W. F. Tolmie, medical officer in 1832 of Hudson's Bay Company of Fort Vancouver, and Archibald Menziès, physician-botanist with Captain George Vancouver's expedition of 1790-95. Youth-on-Age rarely grows east of the Cascades.

FOOD and MEDICINE Makah ate young sprouts raw in early spring. Cowlitz applied a fresh leaf to a boil.

OTHER Many enjoy this as a house plant; it also naturalizes well in woodland gardens. Its purplish flower spike blooms from April; the little new leaves are its most interesting feature.

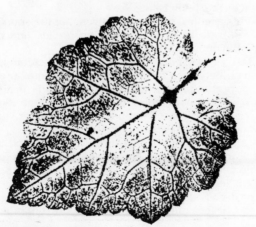

Piggy-back Plant, Youth-on-Age
Tolmiea menziesii

COAST or LODGEPOLE PINE
Pinus contorta   Pinaceae

NAMES *Pinus* is the old Latin name for Pine; *contorta* notes malformation of Coast Pines. Trees twisted by beach winds contrast sharply with those inland, where the trunks are ideal for lodge poles and other uses requiring straight pieces. Adopting different forms in different habitats, this native evergreen was formerly considered two different species. A variety is sometimes added to *P. contorta;* var. *contorta* describes Coast Pine; var. *latifolia* identifies the Lodgepole. Number of needles in a bundle or fascicle helps identify Pines. Two needles are in a bundle on *P. contorta;* they persist from 5 to 9 years. (The only other western Pine with two needles, *P. muricata,* grows in the California Coast range.) Douglas Fir needles are borne singly on the branches. Coast or Lodgepole Pine cones often stay on trees after they have opened and dropped their seeds. A large stand of *P. contorta* may grow where there has been a forest fire, as fire releases seeds for germination.

FOOD Pine cambium was harvested when new needles were growing and pollen cones were ripe; that of Coast Pine is said to develop a taste like turpentine as the season advances, so early harvest is important. *P. contorta* grows in much of British Columbia; its cambium was an almost universal food for natives of the Interior, who usually ate it fresh. Shuswap also dried it for winter use. Southern Oregon Indians ate Pine cambium as a relish or as famine food.

Seeds of nearly all Pines are edible; some are sold commercially. Pine nuts are flavorful, nutritious, and easily digested. Size varies with different species; not all are collected for food. Small seeds of Coast Pine were used for food where larger seeds were not available. Harrington says that seeds of *P. edulis* contain 14.6 percent protein, 61.9 percent fat, and 17.3 percent carbohydrate.

Lewis says Pine needles, species not given, are the source of a F.D.A. approved sugar substitute used in flavoring breads and cakes which intensifies sugar flavor 30 to 300 times.

MATERIALS Uses of *P. contorta* reflect its size, seldom large enough for boards; it was widely used for posts, poles, and mine props, and has been used for paper pulp. Southern Oregon Indians pushed dugouts through shallow waters with poles of Lodgepole Pine. Some Oregon natives made baskets of Coast Pine bark especially for Huckleberries, covering berries with large leaves to keep them fresh.

MEDICINE Quinault treated open sores with pitch and chewed buds for sore throat; Kootenay thought the cambium good for consumption, while Carrier considered cambium helpful for colds. Some groups treated boils with Pine gum.

OTHER Coast or Lodgepole Pine is a pleasing ornamental; it may grow from 40 to 100 feet tall, and mature trees reach an age of 100 to 200 years.

PONDEROSA or YELLOW PINE
Pinus ponderosa   Pinaceae

NAMES   *Ponderosa* means heavy in Latin. This grows abundantly east of the Cascades, where the drier climate is more suited to its needs. *Ponderosa* usually has three needles to a bundle; these last about three years. Cones do not remain long on the branches. Ponderosa Pine may live five hundred years.

FOOD   Some Southern Oregon Indians ate cambium of Ponderosa Pine as a relish or as famine food. Thompson, Okanagan, and Southern Shuswap ate cambium, as did Flathead Salish of Montana, who reportedly used this species more than Lodgepole Pine, collecting it early in the spring from young trees, preferably those which had not yet borne cones. One process of preparing cambium involved baking it in a pit, then smoking it over a rack; it was sometimes wrapped in Skunk Cabbage leaves for storage. Lillooet, Thompson, Okanagan and Southern Shuswap ate the small seeds of Ponderosa Pine raw, or crushed and made into bread. Tea has been made from needles of *P. ponderosa* and *P. edulis*. Resinous taste of needles and buds of Ponderosa Pine, toxic to livestock, usually discourages quantity consumption; eating needles of *P. ponderosa* is particularly dangerous for cattle.

MATERIALS   California Indians made dugouts from fire-hollowed Pines; sometimes twisting sticks of fire starters were made of Yellow Pine; Pine cones are useful for making a quick fire. Ponderosa Pine is second to Douglas Fir in volume of saw timber.

MEDICINE   Pines provide valuable industrial and medicinal products. Primary sources are species from other areas, although our Pines may have similar factors in smaller quantities, less available or less desirable forms. Pine Oil is obtained by steam distillation of wood from turpentine-producing species; it has been recognized in the National Formulary for disinfectant and deodorant uses, and used in paints and synthetic resins. Pine Tar, also distilled from various Pines, has been recognized in U. S. and British Pharmacopoeias for internal use as an expectorant and external use for treating skin diseases. Turpentine occurs in oleoresin reservoirs in sapwood of Pines, especially Southern Longleaf Pine, Slash Pine, and Loblolly Pine. Medicinal use of turpentine has fallen into disrepute, but industrial uses remain important.

Herbal medicine has used Pine Oil as rubefacient, diuretic, and irritant. Turpentine has been used in general and veterinary practice as rubefacient, vesicant, and antiseptic; it has been used for a vermifuge for horses and cattle. Herbal medicine has suggested Pine needle tea for a mild diuretic and expectorant, or the same tea for a beverage; inner bark and pitch have been used to provide an expectorant. Dried inner bark of a Pine of the eastern United States, *P. strobus,* formerly included in the National Formulary, was an ingredient in cough syrups.

PINEAPPLE WEED
Matricaria matricarioides   Compositae

NAMES   The scent of dried flower heads of the crushed fresh plant, thought to resemble that of Pineapple, provided the common name. Another name is Rayless Dog Fennel, as the flower heads have only disk flowers. The scientific name of this native annual comes from the Latin *mater*, mother, and *caria*, dear. *Mater* or *matrix* was often used for medicinal plants; this genus was formerly a favorite of herbalists. The suffix -*oides* means like; how this particular species is like other genus members is not clear.

FOOD and MEDICINE   Pineapple Weed is not recommended for food, although one source suggested a beverage tea from flower heads only, as leaves have a disagreeable flavor. Herbalists used flowers of other *Matricaria* species as tonic, carminative, and sedative. "Mother" in the name may reflect former use to treat infections of the uterus.

OTHER   Flathead of Montana used whole plants in skin bags to keep bugs off meat or berries; dried leaves were sprinkled over meat and fruit for the same purpose. Blackfoot used dried flowers for insect repellent. Kootenay hung flowers in their houses for the fragrance.

COMMON or BROAD-LEAVED PLANTAIN
Plantago major   Plantaginaceae
LANCELEAF or ENGLISH PLANTAIN
P. lanceolata   Plantaginaceae

NAMES   *Plantago*, from the Latin *planta*, sole, plus *agere*, to carry or bear, points out resemblance of leaf shape of some species to the sole of a foot. *Major*, greater or larger, describes the broad leaves of the Common Plantain, *lanceolata*, the lance-shaped leaves of English Plantain. These Eurasian perennials are much-disliked lawn weeds; English Plantain is the more obnoxious. Their size is influenced by the quality of growing conditions. These bloom much of the year in roadsides and disturbed sites. Pollen of Plantain causes hay fever; *P. lanceolata* in particular releases great quantities.

FOOD   Some wildfood authors recommend early spring leaves of Plantains for potherbs, but even these seem quite fibrous and tough. Leaves of *P. major* are said to be better than those of *P. lanceolata*. Plantain leaves have been used for a beverage tea. Seeds can be ground into flour or meal; grain-like heads topping the tall flowering stalks provide a plentiful supply. Most small birds relish the seeds, which have been sold as bird food.

MEDICINE   Early herbalists used these plants Linnaeus knew. Traditional and herbal medicine have considered them refrigerant, diuretic, laxative, alterative, and astringent; leaves have been applied locally to stop hemorrhages. For toothache relief in Europe, fibrous stems of *P. major* were put in the ear on the side of the aching tooth. Herbal medicines combining several plants often included Plantains. The 18th-century European botanist Boerhaave used Plantain leaves to

Broad-leaved Plantain
Plantago major

Lanceleaf Plantain
Plantago lanceolata

relieve pain of aching feet. Fresh leaves of both species, mashed to a pulp or chewed, provided a treatment for cuts, scratches, wounds, Nettle stings and insect bites, including bee stings.

OTHER   Plantain seeds are covered with a coat of mucilage which separates out in hot water; this was at one time used in France to stiffen woven fabrics, and has been used in paper manufacture.

Psillium seed, recognized in the National Formulary for laxative use, is obtained from *P. psillium* and other Plantain species. *P. lanceolata* seed has been a principal adulterant. Soaking seeds in water is a method of determining quality; those from more desirable species are consistently more bulky.

POISON HEMLOCK
Conium maculatum   Umbelliferae

NAMES   *Conium* is the Latinized form of the Greek *konein*, to whirl around, to become dizzy, an effect of eating this plant. *Maculatum*, spotted, points out brown-purple spots on plant stems; these help identification. This Eurasian native, a weed established in roadside ditches and moist disturbed sites over much of North America, blooms from May. The fernlike greenery, up to ten feet tall, is quite attractive. The only other species in this genus grows in South Africa.

MATERIALS   Klallam women rubbed their bodies with Poison Hemlock roots after bathing, in a charm·use intended to attract a man. Snohomish rubbed roots on fishhooks to remove the scent of fishermen.

MEDICINE   Traditional and herbal medicine formerly used Poison Hemlock for internal and external medicines; seeds were official in the U. S. Pharmacopoeia and leaves in the British Pharmacopoeia. This is the Poison Hemlock used in ancient times to put criminals to death; it was given to Socrates in ancient Athens. It should be avoided or handled with extreme caution. All plant parts contain the toxic principle coniine. The toxic alkaloids affect the central nervous system so severely that death from respiratory failure results from eating an extremely small quantity. This plant provides another reason why careful identification is essential if you are considering food use of wild herbs.

POISON OAK
Rhus diversiloba   Anacardiaceae

NAMES   Source of this genus name may be *rhudd*, Celtic for red, or the Greek *rhus*, for some of this genus have red fruits, or it could refer to the red leaves of some fall foliage, or *rhous*, Greek for Sumach or tannin tree, for its use in tanning. *Diversiloba* describes the variable lobes of the three-part leaves. *R. diversiloba* (or *R. toxicodendron*, or *Toxicodendron diversilobum*), which grows primarily in dry woods and thickets west of the Cascades, is sometimes called Poison Ivy. *R. radicans*, found east of the Cascades, is more commonly so identified. Both are native to North America. Some think lower leaflets of *R. radicans* resemble Ivy. *Radicans* means rooting, for the underground stolons or branches which help spread the plant. *R. diversiloba* has rather rounded leaflets; those of *R. radicans* are pointed. Neither of these *Rhus* species, members of the Sumach family, are Oak or Ivy; Oak belongs to the Beech family, Ivy, the Ginseng family.

MEDICINE   Traditional and herbal medicine formerly used Poison Oak as rubefacient and irritant. Body defenses against Poison Oak increased circulation or reacted in ways which improved the condition being treated, despite unpleasant side effects. This remedy does not appeal to me! Fresh leaves of *R. radicans* were formerly included in the U. S. Pharmacopoeia as irritant. Stuhr reported that Indians used *R. diversiloba* to counteract rattlesnake poisoning. Homeopathic medicine has used Tincture of Rhus for an antirheumatic treatment.

Strong tea provides some relief from blisters of Poison Oak, probably because of drying action of the tannin; washing with a strongly alkaline laundry soap is also recommended.

OTHER   A poisonous plant has similar adverse effects for all. Not everyone is susceptible to dermatitis from Poison Oak, so a meticulous definition considers Poison Oak or Ivy allergens rather than poisons. Problems caused by allergens can be every bit as serious as those caused by poisons. Because of volatile oil in leaves, bark, and fruit, even smoke from fire in which the shrub is burned can cause severe skin irritation, as can touching the plant itself or handling clothes which have come in contact with it.

Some California Indians used Poison Oak for basketry and made a black stain or fabric dye from the fresh juice. The color musk has been obtained from shoots of *R. toxicodendron*. Flowers are an excellent source of honey, which is not poisonous. Birds eat the fruit.

QUEEN ANNE'S LACE, WILD CARROT
Daucus carota   Umbelliferae

NAMES   The flowers of Queen Anne's Lace, (no one knows which Queen Anne), appear. very lacy, particularly from above. This is considered the source of our garden Carrot. The curled shape of old flower heads provides another name: Bird's Nest Plant. *Daukos* was the ancient Greek name for some Parsley family member; *carota* is an ancient Roman name. This biennial European weed is widely spread in open fields and roadsides west of the Cascades. Each flower head has many tiny white flowers and one colorful flower near the center. Wet leaves of Wild Carrot cause dermatitis for some people.

FOOD   Pliny reported the cultivated Carrot was eaten raw or cooked and used as a stomachic in Syria. Some wildfood authors recommend wild roots for a cooked vegetable; these are less palatable than cultivated roots. Wild plants supposedly differ from cultivated in size and taste of roots. Dried roasted roots have been used for a coffee substitute. Carrot roots, nourishing and high in sugar content, are good fodder for cattle and horses. If cattle eat leaves, the milk has an undesirable flavor.

MEDICINE   Traditional and herbal medicine formerly used Queen Anne's Lace, especially the seeds, as carminative, stimulant, diuretic, laxative, and emmenagogue, also to treat ulcers and eczema. Folk medicine has used a tea said to maintain low blood sugar levels in humans. Wild Carrot seeds were official in the U. S. Pharmacopoeia from 1820 to 1882; herbalists also preferred the wild plant.

OTHER   The finely-divided leaves of Wild Carrot were used in ladies' hairdresses for balls and banquets during the reign of James I, King of Great Britain and Ireland, 1603-1625.

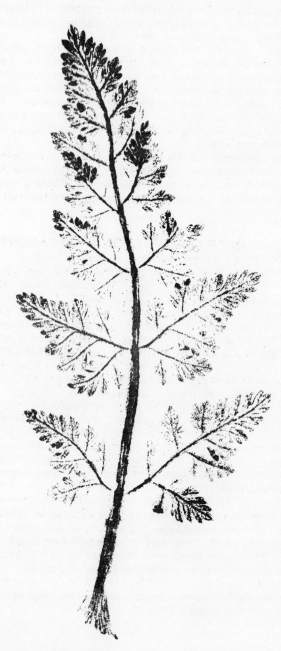

Queen Anne's Lace, Wild Carrot
Daucus carota

WILD RADISH
Rhapanus sativus   Cruciferae

NAMES   Radish is from the Latin *radix, radicis,* a root, especially a Radish. *Rhapanos,* Greek for quick to appear, refers to speedy seed germination; *sativus* means cultivated. This, perhaps originating in Southern Asia or China, has been cultivated for so long that its wild source is now unknown. It was known to and eaten by Egyptians at the time of the Pharoahs. Pinky-purple flowers of this garden escape are common in fields and waste places about April.

FOOD   Northwest wildfood sources do not recommend eating the Wild Radish, which contains a pungent volatile oil. The crisp white root of the cultivated Radish is a good antiscorbutic food; seed pods can be used for pickling.

RATTLESNAKE PLANTAIN
Goodyera oblongifolia   Orchidaceae

NAMES   Mottled green and white leaf markings, supposedly like those of a rattlesnake, gave the common name. *Goodyera* honors a 17th-century English botanist, John Goodyer, 1592-1664. *Oblongifolia,* Latin, describes leaf shape. This ground-hugging native perennial sends up its inconspicuous spike of greenish-white flowers in coniferous woodlands in July or August.

MEDICINE   Klallam women formerly rubbed Rattlesnake Plantain on their bodies as a charm to make their husbands like them better. Cowlitz used leaves for a tonic tea. Herbal medicine has treated scratches, inflammations, and insect bites with dried, crushed and powered basal leaves and roots of Rattlesnake Plantain. Fresh leaves, washed and steeped, have been used for a first aid eyewash.

NOOTKA ROSE
Rosa nutkana   Rosaceae

NAMES   *Rosa* is from the Celtic word for red; Rose flowers and fruit are often red. Common and species names, spelling altered to fit Latin usage, come from Nootka Sound, Vancouver Island, where the species was first identified. This large-flowered Rose, growing mostly west of the Cascades, likes sunny open areas or woodland edges. Nootka Roses often grow along roadside fences, providing shelter and food for small birds and animals as well as pretty flowers.

FOOD   Peoples all over the world have eaten Rose hips. Hips or seed pods of Roses, said to be sweeter after the first frost, are rich in vitamin C, which is not destroyed by cooking. During World War II when citrus fruits were unavailable, the English used Rose hip jam to provide Vitamin C. In addition to Vitamins A and C, hips are a source of iron, calcium and phosphorus, and seeds have Vitamin E. Three Rose hips are supposed to contain as much Vitamin C as one orange. Rose hips are suggested as a good emergency food; they can be eaten raw or cooked, or split and the seeds scraped out and discarded. Rose hips were considered famine food by some Northwest natives, treats for children by others, undesirable for food by still others. Rose hips were sometimes pounded and included in pemmican. *R. nutkana* was widely available in British Columbia; Indians there commonly ate rinds of these Rose hips.

Pulp from Rose hips of all species can be used in jam or jelly; larger, pulpier hips are best. Dried ground hips can be added to soups, stews, or teas. Rose hip products are sold at health food stores for teas, vitamin supplements and food additives. Beverage and medicinal teas have been made from Rose leaves, hips, and flower petals. Some authors suggest adding petals to salads. Peeled spring shoots of wild Roses were used as a potherb by Vancouver Island Salish; some wildfood editors also recommend these.

MATERIALS   Arrow shafts have been made from old straight wood of Roses. Jars of dried Rose petals make a pleasant room freshener.

MEDICINE   Native Americans and herbalists recognized astringent qualities of Roses by medicinal uses. Haida women of British Columbia used peeled young shoots as a tonic and beauty aid; they did not eat these. California Indians made tea from Rose leaves and hips for a beverage or to treat pain. For rheumatism, a tea was made from Rose bark; muscular pains were treated with cooked seeds, and some made a tea from tender roots to treat colds. Rose flowers have been used for centuries in folk medicine. Herbals have suggested a tea from flowers, buds and hips, especially after frost, for diarrhea; a mildly astringent eyewash can be made from Rose buds. A decoction of Rose roots was used as a diuretic. The name Rose fever for hay fever confirms the problems caused by some Rose pollens.

OTHER   Rose Petals, Rose Oil, Rose Water, and Rose Water Ointment have been in the U. S. National Formulary or the U. S. Pharmacopoeia for many years; these were used largely for perfume and for scenting other products. Very small amounts are used in Rose Water products. From 150 to 300 pounds of Rose petals are needed to distill one ounce of Rose Oil. Some species produce more oil than others. Northwest species are not recommended sources, though our Roses undoubtedly contain some of the desirable factors.

The single bright pink flowers of *R. nutkana* make it a nice addition in dry to moist areas of wild gardens both east and west of the Cascades.

CLUSTERED WILD ROSE
Rosa pisocarpa   Rosaceae

NAMES   A characteristic cluster of three to six flowers helps identify this sparingly-armed native, which grows mostly west of the Cascades. Another name is Swamp Rose, for this species often grows in moist areas. *Pisocarpa*, pea-fruited, describes the hips, usually round to pear-shaped. It has straight prickles or spines, 5 to 7 or 9 leaflets on its compound leaves. The small bright pink flowers are seldom solitary; sepals are persistent on the hips.

MEDICINE   Washington Squaxin and Vancouver Island Salish ate hips of this species; Snohomish made a tea from boiled roots for sore throat.

SWEETBRIAR ROSE
Rosa eglanteria (R. rubiginosa)   Rosaceae

NAMES   *Eglanteria*, French for sweet brier, translated in the common name, provides the species, as foliage is said to smell sweet, especially after rain.

OTHER   This, one of the oldest-known and best-loved European species, has escaped west of the Cascades along roadsides and in areas settlers used for meadows. Hips are pear-or egg-shaped; sepals often fall off when the fruit is mature. The somewhat hairy leaflets are rounded at the tips, and have strong-curved prickles. There are one to four pinkish-white flowers on short stalks; leaves have 7 to 9 leaflets 1/2 to 1 inch long, and curved prickles at the stem base.

WOOD ROSE, LITTLE WILD ROSE
Rosa gymnocarpa   Rosaceae

NAMES   Wood Rose identifies habitat and Little Wild Rose points out size of flowers of this native, named by Nuttall in 1840. *Gymnocarpa* means naked fruit in Latin, referring to the hips, from which sepals usually fall early; sepals remain on hips of most Rose species. The name Bald Hip Rose also notes this feature. The Wood Rose, which blooms from May, has 5 to 7 or 9 leaflets; the stem has prickers, not really thorns.

FOOD   These tiny hips were those least used by Northwest Indians. Vancouver Island Salish, Bella Coola and Comox ate the raw outer rinds, but not the seeds.

MATERIALS   Thompson made arrows, handles, and baby carrier hoops of the wood of *R. gymnocarpa*.

OTHER   The Wood Rose is a nice addition to a wild garden.

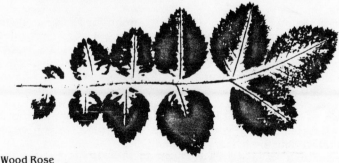

Wood Rose
Rosa gymnocarpa

ST. JOHNSWORT, KLAMATH WEED

Hypericum perforatum   Hypericaceae

NAMES This yellow-flowered Eurasian perennial blooms about June 24, a day dedicated to St. John, so it is St. John's plant. Among the duties of St. John the Baptist was protection from witches, considered especially dangerous on Midsummer Night, June 24. *Hypericum* comes from a Greek verb meaning to hold over in such a way as to protect. When leaves are held to the light, small glands in them seem to be holes, hence *perforatum*, with holes.

MEDICINE The ancients considered this plant useful for a wide variety of internal and external complaints, but now it is scarcely used. Herbal medicine formerly used tops and flowers for an astringent and a diuretic, juice to treat wounds and abrasions and prepare an expectorant, and leaves steeped in oil to treat skin inflammations. Fresh leaves were considered more effective than dried. St. Johnswort causes contact dermatitis in some individuals. Plant parts are poisonous when eaten. Milk of cows which have browsed on it develops an undesirable flavor; most animals will not eat the bitter plant unless they face starvation. Some ingredients result in photosensitization, affecting white areas of animals which have eaten it: a black and white goat may be severely sunburned or have a skin eruption on white spots only. Consumption of more desirable plants, permitting the less appealing to grow, has encouraged spread of this pest. A European beetle, *Chrysolina gemellata*, has been used to combat it.

OTHER California Indians had access to *H. perforatum* early enough to develop a yellow dye from plant buds. A yellow dye has been made from the flowers.

SALAL

Gaultheria shallon   Ericaceae

NAMES Common and species names are variations of the Indian name for this native shrub, described by Pursh. Its creamy-white urn-shaped flowers appear in May. The genus honors Dr. Hughes Jean Gaultier, 1708-86, a French-Canadian naturalist and physician from Quebec.

FOOD Salal, gathered in large quantities by natives of Washington, Oregon, and the British Columbia coast, provided an important, widely-used food. Cakes of mashed and dried Salal berries sometimes weighed ten to fifteen pounds. After soaking, these were often eaten with whale or seal oil. Parts of cakes were used to sweeten other food. Berries also added flavor to soups or boiled roots, some of which were served to William Clark of Lewis and Clark. Wildfood editors recommend berries raw, for pies, sauce, and jam. Eating too many gives a laxative result. Flowers furnish nectar to bees. Berries are an important wildlife food; bears like Salal.

MATERIALS Salal leaves were spread under drying berries, used to line cooking pits or cover food in them. Skunk Cabbage leaves furnished wrapping for stored Salal cakes.

Salal
Gaultheria shallon

Salmonberry
Rubus spectabilis

MEDICINE   Klallam and Quileute spit chewed leaves on sores and burns. Quinault chewed leaves to relieve heartburn and colic. Swinomish and Samish treated coughs and tuberculosis with a leaf tea. Skagit made a mild leaf tea for a convalescent tonic.

OTHER   Makah and Quileute combined dried Salal leaves with Kinnikinnick to use as smoking tobacco. Nootka made a purple dye from Salal berries; Kwakiutl added Black Twinberry for a stronger color.

This broad-leaved evergreen is a nice ground cover in a woodland garden. David Douglas introduced Salal to England, where it became popular as an ornamental. Salal leaves are among forest products gathered for florists. These, known as Lemon Leaves, command a lower price per bunch than Evergreen Huckleberry, which goes farther in floral arrangements.

SALMONBERRY
Rubus spectabilis   Rosaceae

NAMES   Indians identified the April bloom period of this native shrub as the time when salmon came up the river. Watch for these and confirm the accuracy of this observation, one suggested source of the common name. *Rubus*, from the Latin *rubere*, to be red, and *spectabilis*, remarkable or showy, for the bright pink flowers, provide the scientific name. As the season advances, both ripe berries and flowers may be on the bush at the same time. Fruit, salmon colored, varies from red to yellow. Some say lighter-colored berries taste better. Salmonberry bushes grow in moist locations, mainly west of the Cascades.

FOOD   Many Northwest Indian groups ate peeled Salmonberry sprouts in spring, raw or cooked. They were sometimes cooked in fire-heated pits and eaten with dried salmon. Lower Thompson were among coastal British Columbia groups which ate large quantities of sprouts and berries. Fresh berries were eaten wherever they were available throughout Oregon, Washington and British Columbia, but were often considered too soft and watery to dry. They were favorites of children and are eaten now with sugar. Wildfood editors recommend peeled young shoots, raw or cooked berries, dried leaves for a beverage tea, and wine from the fruit.

MATERIALS   Kwakiutl sometimes made arrow shafts from Salmonberry stems; Kwakiutl and Haida made spears for throwing games. Comox added water to covered steam-cooking pits through the hollow stems of Salmonberry.

MEDICINE   Quileute chewed Salmonberry leaves, or bark in winter to spit on burns. Makah laid pounded bark on an aching tooth or festering wound to kill pain. Quinault cleaned infected wounds and burns with a bark solution. To lessen labor pains, Quinault made a bark tea with sea water. Dried bark also provided an astringent tea for treating stomach disorders.

OTHER   This is another shrub recommended for a wild garden. Those who enjoy interesting plant patterns will like the butterfly made by folding back the single unmatched leaflet in the three-part leaf.

SALSIFY, PURPLE or YELLOW OYSTER PLANT
Tragopogon porrifoliua; T. dubius   Compositae

NAMES Salsify may be a variant of *sol,* sun, and *sequens,* following, as flowers open and close early; or perhaps it is from the Latin *saxifricta, saxum,* rock, and *fricta,* rubbed. Goatsbeard, another name for this vegetable-garden escape, translates *Tragopogon,* Greek *tragos,* goat, and *pogon,* beard; the seed has silky fibers or threads. Oyster plant says raw roots and oysters taste alike. *Porrifolius,* leek-leaved, describes leaf shape; reason for *dubius,* doubtful, is unknown. Don't look for blooms, (purple, *T. porrifolius,* yellow, *T. dubius),* after noon, for their tightly-closed buds will be lost in the grasses where they grow.

FOOD Salsify roots are eaten as parsnips, green tops for potherbs, young shoots as asparagus. Some say the purple species is better, because less fibrous. Settlers brought these European biennials to the United States early; Indian uses were noted for the Midwest, but not the Northwest.

MEDICINE Salsify, used medicinally in medieval times, has fallen out of use. Milky plant juice, coagulated, has been chewed for gum and used to treat indigestion.

OTHER Salsify seedheads are large and fluffy, like giant Dandelion heads. Handled carefully and sprayed, these are nice in dry bouquets.

SCOTCH BROOM
Cytisus scoparius   Leguminosae

NAMES Scotch Broom and *scoparius,* Latin for broom, report use of the wiry stems by Scottish ladies. *Cytisus* is a variation of Cythera, an Ionian Island, from which Scotch Broom may (or may not) have come. As there are so few tiny leaves that stems may appear leafless, the green stem is important for photosynthesis.

This southern European shrub with yellow spring blooms, often planted on highway cuts, has been very successful west of the Cascades in spreading to adjacent areas. Scotch Broom grows well in poor soil as its deep roots help it reach water; roots of this and other legumes are able to fix nitrogen. Ripe seedpods split with explosive force and surprising noise, sometimes throwing seeds more than twenty feet. Ants which eat oil on the seeds also scatter them. The heat-resistant seeds often survive a fire.

FOOD Broom tops were used to flavor beer before Hops were introduced. Young flower buds have been pickled as a substitute for capers. Current wildfood editors suggest roasted seeds as a coffee substitute. Do not eat uncooked plant parts; the toxic alkaloids in Scotch Broom are apparently destroyed by heat. These also poison animals. The abundant pollen attracts bees.

MEDICINE Use of *C. scoparius* is known from early Anglo-Saxon use, Welsh physicians of the early Middle Ages, and herbalists of the fifteenth century. More recent herbalists have used powdered dried

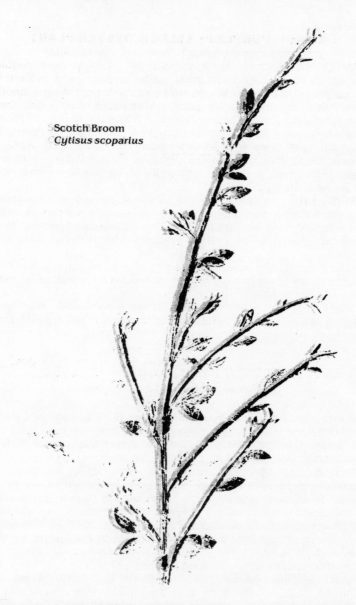

Scotch Broom
Cytisus scoparius

tops, dried flowers, and sometimes seeds to prepare a cathartic and a diuretic tea. Dried Broom tops were formerly official in the U. S. Pharmacopoeia as a cardiac stimulant inferior to *Digitalis*, also as narcotic, cathartic and diuretic; it was considered poisonous in large doses. Broom juice has been official in British, French, German, and United States Pharmacopoeias.

OTHER The strong fibers have been used in paper and cloth. Because of its tannin content, bark has been used in tanning leather. A green dye has been made from young tops and leaves. Cultivated Brooms are better for home gardens than *C. scoparius.*

Scotch Broom was one of three plants used to stabilize drifting dunes which were burying homes, roads, and forests at the northern Oregon coast. In a project developed by the U. S. Soil Conservation Service in 1934, five square miles were planted over a ten-year period. European Beach Grass and native Coast Pine, *P. contorta,* also helped remedy damage caused at least in part by earlier heavy use by livestock. This outstanding example of dune stabilization unfortunately phased out much native vegetation, although some had been endangered by the spreading dunes.

Gorse, *Ulex europaeus,* a related yellow-flowering shrub also escaped from cultivation, has thorns or spines and a great deal of oil in its branches, making this European native a serious fire hazard. It was a factor in the 1936 fire at Bandon, on the Oregon coast.

SCOURING RUSH
Equisetum hyemale   Equisetaceae

NAMES This scours effectively. *Hyemale,* of winter, recognizes it as an evergreen; its stalks persist for several seasons. Reproductive parts of Scouring Rush grow at the tip of the unbranched green stalks instead of in a separate short-lived form as in Common Horsetail.

MATERIALS Historically, cabinet makers used Scouring Rush rather than sandpaper. Skokomish, Cowlitz and Quinault smoothed arrows and other artifacts with Scouring Rush. Pioneers appreciated this use, effective because *Equisetum* species collect silica (sand) from the soil. Try this on your camp cooking pots! Quileute, Swinomish and Cowlitz used the black root for design in coiled baskets.

MEDICINE Cowlitz washed vermin-infested hair with a decoction from stalks; Quinault used a similar solution to treat female problems. Quileute rubbed their bodies with *E. hyemale* to feel strong. Early herbalists noted astringent uses. Gerard, herbalist of 1633, recommended applying crushed Scouring Rush to wounds; modern herbalists agree. Tonic and diuretic uses have also been reported. These species contain aconitic acid, which poisons horses when included in hay, though cattle apparently eat fresh plants without harm.

OTHER Scouring Rush has recently been suggested for wet gardens. *E. hyemale* is desirable; Common Horsetail, *E. arvense,* is a pest.

SEDGES
Carex and *Scirpus* species   Cyperaceae

NAMES Middle English *segge,* sedge, *Carex,* Latin for sedge, *scirpus,* Latin name for some of the species. An old jingle says Sedges have edges. Feel at the base of grass-like clusters; if the leaf bases are triangular, it is a Sedge. Grass stems are round. Sedges, an interesting part of our flora, grow in moist areas and have inconspicuous flowers. "Tules" are Sedges. Let knowledgeable individuals identify the species, but do be aware of this group of plants.

FOOD and MATERIALS The grasslike leaves of Sedges were extensively used by Indians of the Pacific Northwest for baskets and hats. Dried seeds of *Scirpus* species were used for food.

SERVICE or SASKATOON BERRY
Amelancier alnifolia   Rosaceae

NAMES Former classification in the genus *Sorbus* makes Sarvis and Service Berry logical. Saskatoon Berry is often used in Canada; it comes from the Blackfoot name. *Amelancier* is the French name for a cultivated Hawthorn; *alnifolia,* Alder-leaved, suggests leaves resemble those of Alder. Other names are June Berry, for ripening time, and Oso Berry; *Oso* is Spanish for bear. Common plants often have several names; this has to be widely available.

FOOD Service Berry was a staple of Indian diet where abundant. This species is extremely variable; flavor and size of berries from different bushes or areas are often dissimilar. Coastal fruit is reported inferior to that from farther east and north, also more subject to insect attack.

Lewis and Clark told of bread made by combining Service Berries with pounded seeds of Balsam Root and Pigweed. Indians ate berries raw and cooked, and dried them in large loaves for later use. Chehalis of Washington, and California groups used dried berries to season meat and soups made with roots. Some pounded Service Berries with meat and animal fat to make pemmican. Oregon Indians made a beverage tea from the leaves. Current wildfood books recommend berries fresh, for jellies, preserves, pies, and wine, and dried as a substitute for raisins or currants. Deer and elk relish young twigs and foliage; many birds and animals feed on the small, apple-shaped fruits. Young bushes do not survive if overgrazed.

MATERIALS Some Indians made arrowshafts from Service Berry shoots. Samish and Swinomish made spreaders for halibut rigging from the tough wood. Snohomish used the wood for disks for the gambling game.

MEDICINE California Indians boiled green inner bark of Service Berries to prepare an eyewash.

Service Berry
Amelancier alnifolia
upper section of leaf notched

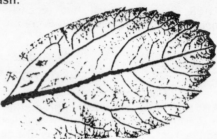

Sheep Sorrel
Rumex acetosella

SHEEP SORREL or SOURGRASS
Rumex acetosella   Polygonaceae

NAMES *Rumex*, an ancient Latin name, is from *rumo*, to suck, for Romans sucked these leaves to lessen thirst; *acetosella*, diminutive form of the Latin *acetum*, vinegar, says this little vinegar plant is smaller or less sour than Garden Sorrel, *R. acetosa*.

FOOD Those who enjoy Sourgrass as a tasty nibble, for salad, or as a potherb, help eliminate a troublesome weed. Wildfood authors suggest eating limited quantities because of the high oxalic acid content. This European perennial has inconspicuous blooms from May. See Violets

MEDICINE Herbal medicine recommends the fresh herb for a refrigerant and diuretic; *R. acetosa*, Garden Sorrel, has similar uses. Remember this for Nettle stings; see also species comments under Dock

OTHER This plant has been known for a long time; Linnaeus used a name familiar to his contemporaries. It grows well in acid soil; creeping rootstocks help it spread. A 1786 dye list reports roots gave a musk color.

SHEPHERD'S PURSE
Capsella bursa pastoris   Cruciferae

NAMES   The common and scientific name are the same: *capsella,* diminutive of Latin *capsa,* box, and *pastoris,* little box that is purse of shepherd, from the shape of the seedpod. This European annual was introduced early enough to develop food uses by Indians of the Midwest and California.

FOOD   Wildfood authors recommend stems and leaves of young plants for potherbs and salad, and seeds parched and ground into flour. The flavor is said to be similar to that of cabbage but more delicate.

MEDICINE   Herbal medicine has used the dried plant as diuretic, stimulant, antiscorbutic, astringent, for stopping hemorrhages, and to clean wounds. Traditional medicine at one time recommended fresh juice or an extract from dried plants of Shepherd's Purse for similar uses.

SKUNK CABBAGE
Lysichitum americanum   Araceae

NAMES   Skunk Cabbage refers to the smell, which is not so bad if kept in the Great Outdoors; the odor attracts pollinating insects. *Lysis,* Greek for loosing, and *chiton,* a tunic, describe the spathe or modified leaf surrounding the flower stalk; this perennial is a native of North America. The bright yellow blooms are conspicuous in February and March in boggy places; bees like the pollen. There are only two species in this genus; the other grows in northeast Asia.

FOOD   Some Indians ate Skunk Cabbage, though Mohney says it was never a staple of Indian diet, even where widely available. Quinault roasted the white underground part of the leaf stem. Tillamook, Lower Chinook, and Molalla ate the roots. Young leaves have served as potherbs, roots roasted in pits or dried and ground into flour. Preparation requires special care because the plant contains the toxic calcium oxalate. Calcium oxalate causes mineral imbalance in the blood; its small crystals are intensely irritating to tender tissues of mouth and tongue. Skunk Cabbage is related to the Polynesian Taro which also requires treatment to remove toxins. Native Americans ate available food, and what they learned to eat might well give us discomfort. Bear and elk are said to eat the roots and to tear up swampy areas to get them. Some Arum family plants have extremely acrid juice, some poisonous, possibly due to calcium oxalate or a saponin. Boiling is said to destroy irritant properties, but why push your luck?

MATERIALS   These leaves, largest of any in our area, served in ways we use paper towels. They helped Northwest peoples cover food and cooking pits, provide a clean surface for preparing food, line berry baskets. Use to cover food in cooking pits apparently did not affect food flavor. Folded leaves fastened with twigs could make a drinking cup or a basket for berry picking. Molalla made a yellow face powder from the dried and powdered flowers and the spathes or large bracts which enclose the inflorescence.

MEDICINE   Skokomish soaked roots in water for a physic. Makah boiled roots for a blood purifier; Quileute drank the same liquid to bring about easy delivery; Kwakiutl, Quileute and Skokomish used a poultice of leaves or roots on cuts and swellings. Makah treated pain with warmed leaves. Stuhr says the root of the native Skunk Cabbage was the chief ingredient of the patent medicine "Skookim," reported as stimulant, antispasmodic, and emetic, for use as a salve for ringworm, swellings, and inflammatory rheumatism. These are similar to uses for which the eastern Jack in the Pulpit, also an Arum family member, was included in the U. S. Pharmacopoeia from 1820 to 1882. Dennis lists three garden and house plants in the Arum family which are toxic if eaten.

OTHER   Big leaves and handsome flower spikes of Skunk Cabbage could add interest to wet areas in wild gardens; it is sold in some nurseries. Some recommend flower stalks for cut flowers; they are said to be free of the strong odor, which is supposedly concentrated in the leaves; I haven't tried this!

SMALL TOOTHWORT, SPRING BEAUTY
Cardamine pulcherrima (Dentaria tenella)   Cruciferae

NAMES   Spring Beauty agrees with the species name *pulcherrima*, very beautiful. Other plants are also called Spring Beauty; see p. 25 Someone, sometime, thought tubers of some species resembled teeth, hence Tooth Plant. (*Dens* of the synonym *Dentaria* means tooth in Latin.) *Cardamine*, (Greek *kardia*, the heart, and *damao*, to strengthen), describes a medicinal use for another species. The name Pepperwort says the edible roots taste hot or spicy. Scattered flowers on the forest floor in February and March, mostly west of the Cascades, are a welcome sign of spring.

FOOD   Dioscorides included a similar plant in *De materia medica*. Current wildfood authors recommend this native, but sources consulted did not mention food use by Northwest Indians. Raw leaves, stems and roots are said to add a pleasantly peppery taste to salads. I have not seen them growing in large enough quantities to be comfortable about harvesting them for food.

OTHER   Small Toothwort grows well in rich soil in partially shady locations in wild gardens.

SNOWBERRY or WAXBERRY
Symphoricarpos albus (S. racemosus)   Caprifoliaceae

NAMES   Snowberry refers to the waxy-white berries, which hang on through most winters. *Symphoricarpos, (syn,* together, *phorein,* to bear, and *karpos,* fruit), describes the clusters of white, *albus,* berries. The pinkish-white urn-shaped flowers bloom in May.

Snowberry
Symphoricarpos albus

FOOD   At least one author says the berry is bitter; Kirk says it is insipid but edible, raw or cooked. Squaxin, a Salish group of Washington state, dried them for winter use; Sturtevant quotes an 1870 U. S. Dept. of Agriculture report saying Oregon and Washington Indians ate them. British Columbia groups considered them inedible or even poisonous. If they were highly regarded as food for man or beast, they surely would not stay on the bushes so long!

MATERIALS   Native Americans made arrowshafts and pipestems from Snowberry stems. Deer and elk browse the leaves.

MEDICINE   Chehalis rubbed berries on the hair as a soap, applied bruised leaves as a poultice for cuts, spit chewed leaves on cuts, or washed them with a leaf decoction. Green River disinfected festering sores with the plant. Skagit treated tuberculosis with a bark solution. Skagit ate berries as an antidote for poisoning, confirming emetic and cathartic qualities noted by herbalists. Klallam boiled leaves for a cold cure. Sweet said Indians steeped pounded roots for a decoction to treat colds and stomachache; he reported the poison saponin is in leaves only. The Latin *sapo, saponis,* means soap; plants containing saponins

152

produce a soapy lather and were often used for soap. Dennis notes one reported case of poisoning of children who ate Snowberries. Muenscher does not list any *Symphoricarpos* species.

OTHER   Snowberry provides shelter for birds and small animals, and is a fairly important honey source. It adds interest to wild gardens. Meriwether Lewis brought this shrub to the east coast in 1805.

Sow Thistle
Sonchos oleraceus

SOW THISTLE—COMMON, FIELD, or PRICKLY

Sonchus oleraceus, S. arvensis, S. asper   Compositae

NAMES   *Sonchus*, (Greek *sonchos*, hollow), recognizes the hollow stem of Sow Thistle, a name acquired because pigs, (sows, that is), relish them. Rabbits like them, too. *Oleraceus* comes from *oleraceous*, edible. *Asper* means rough; Prickly Sow Thistle, *S. asper*, is usually more prickly than the similar Common Sow Thistle, *S. oleraceus*. The Field Sow Thistle, *S. arvensis*, of the field, a more troublesome farm weed, spreads from underground stems. All are undesirable, but *S. oleraceus* may be the best known in gardens.

153

FOOD   Peoples of many times and many parts of the world have eaten Sow Thistles. The ancients considered the perennial *S. arvensis* wholesome and strengthening. It is suggested for salads and potherbs, as are *S. oleraceus* and *S. asper,* both annuals. Changing cooking water helps reduce bitterness. These European weeds grow in open fields and roadsides.

MEDICINE   Herbalists have used the milky juice of leaves and stems of *S. oleraceus* as a diuretic.

AMERICAN SPEEDWELL, BROOKLIME
Veronica americana   Scrophulariaceae

NAMES   Speedwell is an old English benediction for departing guests. Brooklime, from the middle English *brok,* brook, and *lemeke,* a kind of plant, refers to a common habitat. *Veronica* honors St. Veronica.

FOOD   Stems of these native North American wetland perennials are used for salads and potherbs. The little blue flowers are attractive from May.

MEDICINE   Herbal and traditional medicine formerly used leaves and flowers of several *Veronica* species as diaphoretic, antiscorbutic, expectorant, and diuretic, though it is no longer recommended.

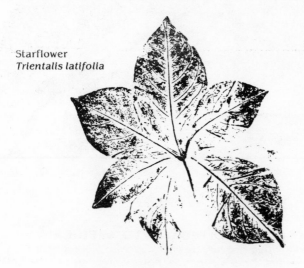

Starflower
Trientalis latifolia

STARFLOWER, WOODLAND STAR
Trientalis latifolia   Primulaceae

NAMES   Woodland Star notes habitat and flower shape. *Trientalis* translates as third of a foot, about the height of these native perennials, which grow chiefly west of the Cascades. The small pinkish-white stars above a whorl of 3 to 7 broad leaves, *latifolia,* are welcome volunteers in my yard. They naturalize well in shady wild gardens. Western or Broad-leaved Starflower are names used for this genus, which has only three species.

FOOD   Some wildflower authors call this Indian Potato, and say it is abundant, starchy, and nutritious. Although food sources consulted did not mention this use, the name should mean something.

MEDICINE   Cowlitz squeezed juice of Starflower in water for an eyewash.

STINGING NETTLE
Urtica dioica, (U. lyallii)   Urticaceae

NAMES   The Anglo-Saxon name for this plant, *netel,* may have come from the Dutch *noedl* or needle, for the tiny hollow hairs pierce the skin like a needle before breaking to release their formic acid content. Another suggested root word is *nere,* to sew, because the plant provided sewing thread. *Urtica,* from the Latin *urere,* to burn, refers to the sting; *dioica,* two houses, indicates a dioecious species, having male and female flowers on separate plants. Many Northwest Nettles are monoecious and have both male and female flowers on the same plant. See Plant Names, p.

FOOD   For hundreds of years, *U. dioica,* native over much of the northern hemisphere, has supplied food, materials, and medicine. Young leaves, picked when about six or seven inches tall, are good vegetables; cooking destroys the sting. If you wish to eat·uncooked Nettles, roll the leaves tightly and squeeze them; the hairs will be broken and unable to do their hypodermic needle thing. Prepared as Spinach or other greens, perhaps with bacon or lemon juice, Nettles are a tasty dish. For a cream soup, add pureed cooked greens to a chicken base, as in making a creamed Spinach soup. Settlers looked forward to this spring vegetable, high in vitamin C, iron, and protein. Dried leaves supply a beverage or a medicinal tea. Fresh Nettle leaves boiled in salted water provided a substitute for rennet in making cheese; a Nettle beer was made for many years. Gloves are suggested for handling Nettles; a heavy plastic bag is an effective substitute.

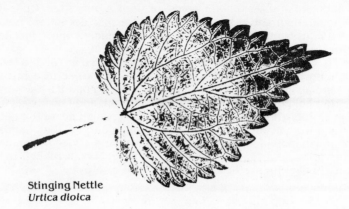

Stinging Nettle
Urtica dioica

The name Indian Spinach says something, though Gunther's study did not report food use. Saanich, Cowichan, Sechelt and Haida, British Columbia Coastal groups, ate leaves and stems in the spring. South Okanagon and Upper Lillooet, Interior groups, boiled and ate Nettles.

MATERIALS   Nettles were a major fabric source for Germanic and Scandinavian nations during the Middle Ages, still used in the 16th and 17th centuries in Scotland. Linen fibers are easier to process than Nettles, so use declined after the introduction of flax.

Germans were short of cotton during World War I and decided to increase fabric supplies by processing Nettles. When quantities of wild fiber proved insufficient, Nettles were cultivated in Germany and Austria. Approximately ninety pounds were needed for a single shirt, though sugar, starch, protein and ethyl alcohol were desirable by-products. Rich deep soil essential for good quality fiber could produce other important crops at a lower cost, so Nettle cultivation was less helpful than had been anticipated.

Northwest Indians made a highly-valued two-ply cord from Nettle fibers, using it for tying, binding, tumplines, snares, harpoons, fishing lines, fish, duck, and deer nets. British Columbia Indian women made twine, though Bella Coola men made nets. Saanich of Vancouver Island spun Nettle fiber with bird down to make blankets before mountain goat wool was easily available.

Nettles have been grown in Russia and Sweden as a fodder crop, cut several times a year. Nettles are an indication of rich soil; they are valuable as fodder when included in hay; feeding chopped Nettles to chickens is said to increase laying capacity. Nettle fibers have been used in Russia and France for making paper. I have included Nettles in handmade paper, but not for the basic fiber, as home-processed fibers are rather coarse. The strong fiber of Ramie, *Boehmeria nivea,* an eastern Asian plant of the Nettle family, is still used in clothing.

MEDICINE   Whipping with Nettles, called urtication, improves muscle tone by increasing blood circulation. Quileute rubbed bodies of seal hunters with Nettles before they set out to hunt seal. This helped them stay awake to keep in sight of land so they could find their way home. Chehalis and Quileute whipped rheumatism sufferers with Nettles; Snohomish used this for colds, or used a Nettle tea. Quileute treated rheumatism and other ailments with a root decoction. Haida cooked Nettle rhizomes as medicine; Gitksan made a medicinal tea from Nettle leaves.

Dr. W. F. Tolmie of Hudson's Bay Company decided in 1835 to treat scurvy among his men with Nettles after noting its use in a British Columbia Indian village. Chehalis and Skokomish prepared a hair wash for girls from boiled Nettle Roots. Squaxin, Lummi, and Cowlitz used a leaf tea in childbirth; it was thought the Nettles would scare the baby out. Quinault women chewed tips of Nettle plants during labor. Samish and Swinomish pounded Nettle plants and Grand Fir together and boiled this to make a tonic solution for bathing. Makah rubbed their bodies with Nettles for purification after handling corpses.

Herbal medicine has used Nettles as astringent, blood purifier, tonic, and diuretic, for a hair rinse after shampoo, a hair tonic, scalp conditioner, and growth stimulant. The dried herb, roots, seeds, or fresh leaves, have been included in preparations to stop internal bleeding and to treat burns. Homeopathic medicine has treated skin complaints with Stinging Nettle products. Traditional medicine formerly recognized use of Nettle herbs externally as an irritant, and internally to treat hemorrhages. The ancients applied Nettles as a counterirritant to excite activity in paralyzed limbs. Roman soldiers brought to the outpost of England a Nettle species which still grows near their former settlements.

Stinging Nettle pollens are considered a cause of hay fever; it causes contact dermatitis, as many can testify. Nettle juice is one antidote. To lessen the pain of stings, herbalists suggest rubbing with leaves of any of five weeds growing near the plant. Mentioned are Elderberry, Curly Dock, Broad-leaved or Lance-leaved Plantains, Bracken Fern, Thimbleberry, fuzz from fiddleheads of young ferns—or whatever may be handy. *Rumex* species—Docks—are considered especially helpful. Chewing these leaves to provide interaction with saliva increases the helpful effect; for best results, juice or crushed leaves should be kept on stings up to half an hour. Rubbing alcohol also alleviates pain of stings.

OTHER   A green dye has been made from the Nettle plant, which contains tannin; from roots with an alum mordant, a yellow dye. Tlingit made a red dye from Nettles boiled in urine. Caterpillars of Red Admiral, (*Vanessa atalanta*), Painted Lady, (*Cynthia cardui*), and West Coast Lady, (*Cynthia anabella*), butterflies eat Nettle leaves; it is a major food for the Red Admiral.

Stinging Nettle is an extremely interesting, widely-available plant; it has had many different uses. I try to keep a respectful distance!

WILD STRAWBERRY
Fragaria vesca (F. bracteata)   Rosaceae
NAMES   Strawberry may be from an old Anglo-Saxon verb meaning to strew, as leaves are strewn about wide areas. *Fraga*, from the Latin verb *fragare*, to emit fragrance, reports sweet-smelling fruit. *Vesca* means weak or thin; leaves of this woodland species are thinner than those of plants growing in more exposed sites. White flowers in May precede the little red berries which small animals and patient people enjoy in June and July.

FOOD   Makah, Quinault, Chehalis, Klallam, Squaxin, Puyallup, Nisqually, Skokomish, Lower Chinook, Swinomish, Oregon Indians, and most British Columbia coastal and interior groups ate these, most fresh, some also dried. Quinault considered them party food. Skokomish ate them as a treat while walking. Wild berries have better flavor than those of cultivated varieties. Strawberries are rich in vitamins A and C. Leaves are rich in C, and can be used fresh or dried; two large handfuls of young leaves with a quart of boiling water make a

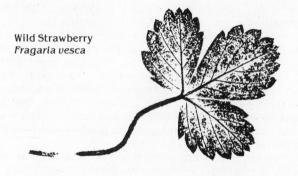

Wild Strawberry
Fragaria vesca

beverage tea. Some Indians added Thimbleberry or native Wild Blackberry leaves for tea. Thompson sometimes used Wild Strawberry flowers, leaves, and stems to flavor roots in cooking pits. Ruffed grouse, robins, black and grizzly bears, small rodents and other wildlife eat the berries.

MEDICINE Quileute spit chewed leaves on burns. Skokomish boiled the whole plant for a diarrhea treatment. Herbal medicine used mostly leaves, sometimes fruit or roots, as laxative, diuretic, and astringent, or to treat diarrhea. Traditional medicine suggested leaves as an astringent for diarrhea, and root tea as a diuretic. Herbalists have used Strawberries to remove tooth discoloration; fresh fruit juice left on teeth about five minutes, then washed off with baking soda is said to be effective. Juice of a fresh Strawberry rubbed on the face after washing whitens the skin, and a similar treatment is helpful for sunburn. Leaves, roots, fruit and fruit juice of *F. vesca* have been used in England for gargles and lotions and to treat oral infections.

OTHER The Beach Strawberry, *F. chiloensis,* growing only in coastal areas, is among the parents of most cultivated Strawberries. Wild Strawberries are good ground cover in a native garden; *F. chiloensis* does not spread aggressively. Roots of *F. vesca* were included in a 1786 list of dye plants; color, cinnamon.

WESTERN SWEET CICELY
Osmorhiza chilensis   Umbelliferae
NAMES Cicely may be a corruption of *Seseli,* an ancient Greek term for some fragrant plant. *Osme,* Greek for scent, *rhiza,* root, and *chilensis,* say a sweet-scented root from Chile. This native perennial blooms in wooded areas in April. Its licorice flavored seeds, efficient hitchhikers, are treats for passersby in May and June.

FOOD Thompson and Lower Lillooet ate roots, raw or cooked, early in the season. Wildfood authors suggest the anise-flavored roots, fresh or dried, as seasoning. Flavor strength varies in different species.

Sweet Cicely
Osmorhiza chilensis

MEDICINE Swinomish chewed the root as a love charm. Some Indians used seeds for seasoning, tea from roots for colds or pneumonia, and raw leaves, chewed, for a physic.

OTHER The ferny foliage of Sweet Cicely is attractive in a wild garden; seed stalks are delicate additions to winter bouquets. As with any seeds in dried arrangements, a spray helps prevent dropping.

SWORD FERN
Polystichum munitum   Polypodiaceae

NAMES Sword Fern reports an opinion that a single leaflet or pinna resembles a broadsword; these leaflets are attached to the stem at the center vein only. *Poly,* Greek for many, and *stichos,* rows, report many rows of spores. *Munitum,* (Latin *munire,* to fortify), says the pinnae have sharp edges. The large evergreen plants are attractive in coniferous woods the year around.

FOOD The edible rhizomes have been roasted, steamed or boiled. Quileute and Quinault baked rhizomes in pits; Makah and Klallam boiled them. Some British Columbia groups considered Sword Fern roots starvation food. Wildfood authors may not recommend these, as they are hard to dig and clean. They are probably best for emergency food.

MATERIALS Sword Fern leaves were used to line cooking pits, cover cooking food, wipe and cover uncooked food. Cowlitz tied Sword Fern leaves together with Maple bark to use for mattresses.

MEDICINE Cowlitz washed sores and treated sore throats with a root preparation. Quinault boiled roots for a hair wash or dandruff treatment. Swinomish chewed and swallowed curled young leaves for sore throat or tonsilitis. Quinault scraped spore sacks from leaves and put these on burns. Lummi women chewed the unfolding leaves or fiddleheads to help in childbirth. Willamette Valley Indians made a tea from the rhizomes to ease pain, and a drink of boiled stems to ease labor.

Sword Fern
Polystichum munitum
tip of frond

OTHER These, like other ferns, reproduce by spores. The prothallium, on which the reproductive parts are formed may be about 1/2 inch in size. (Greek, *pro*, for, *thallein*, to sprout). The young fern develops so slowly that it may not be recognizable as a Sword Fern until it is as much as five years old. Sword Fern provides a multimillion dollar business for florists. Up to a fourth of the fronds can be removed without endangering the plant. Licensed collectors gather ferns, some of which are shipped east in carload lots. Funeral sprays provided ninety per cent of floral use of Sword Fern in 1961.

TANSY RAGWORT
Senecio jacobaea   Compositae

NAMES Fringed Tansy-like leaves gave the name Tansy; rag fits the look of this wort or plant. *Senecio*, from *senex*, an old man, describes the seedheads and hoary appearance of some species; *jacobaea*, for St. James: *Jacob* is the Latin form for James. This European perennial or biennial blooms from July, spreads from its roots, and grows new plants from damaged roots. It has become a serious weed west of the Cascades since the early 1960s. Many of us have watched Tansy Ragwort increase in waste land and along roads, hastened by absence of its native enemies. It does not grow well in the shade. As with other aggressive introductions, it crowds out more desirable natives.

MEDICINE Herbal medicine formerly employed Tansy Ragwort, leaves as emollient poultices, plant juice as a wash for burns and sores, and to treat bee stings, and a root decoction for inward wounds. At the turn of the century traditional medicine used leaves as a bitter, antirheumatic, vulnerary, emmenagogue, and antispasmodic. Its alkaloid content damages the liver, and use has been largely discontinued. See Groundsel.

OTHER The strong undesirable taste of Tansy Ragwort discourages livestock from eating it, thus helping it spread. Its cumulative poison causes liver damage to cattle and horses. Some handweavers raise sheep, less susceptible to these poisons, for the dual purpose of controlling Tansy Ragwort and growing wool to spin their own yarn. The chopped whole plant makes a green dye; flowers with alum make a yellow dye.

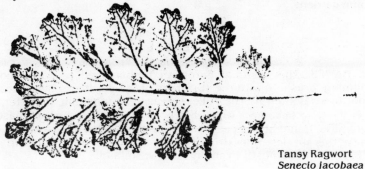

Tansy Ragwort
Senecio jacobaea

This weed in Oregon alone costs more than five million dollars annually in control agents and livestock losses. Three types of biological control have been used, the Cinnabar moth, an Italian flea beetle, and the seed fly. Best known of these is the Cinnabar moth, *Tyria jacobaea*, from northern Europe. This handsome red and black moth has been introduced in many Northwest locations, including all Oregon counties. In summer its larvae help control by eating the flowers. Life cycle of the moth and blooming cycle of the plant do not always coincide, so other methods of eradication are important. The flea beetle is released in the fall; its larvae eat stems and roots during winter and spring. The seed fly eats developing seeds in the summer.

The Italian flea beetle was very effective at the Nature Conservancy preserve at Cascade Head on the Oregon coast, as indicated by a two-year study reported in 1984. The Cinnabar moth larvae were less helpful, but did reduce the size of the stand. Grazing with sheep, burning, mowing, and hand pulling were also considered or tried at Cascade Head, but were less satisfactory. Oregon State University scientists and extension specialists report some progress in the past five years in reducing density through use of biological and chemical controls (herbicides).

TANSY
Tanacetum vulgare   Compositae
NAMES *Tanacetum,* (Greek *tanaos,* long or large, and *akeomai,* to heal), reflect belief that Tansy had healing power; or it may be from the Greek *athaton,* immortal, because the flowers last so long. *Vulgare* is Latin for common. This Eurasian perennial, one of the medieval strewing herbs, escaped from gardens.

MEDICINE Herbal and traditional medicine have used leaves and tops as diuretic, stimulant, vulnerary, emmenagogue, diaphoretic, and febrifuge, seeds as a vermifuge. Tops and leaves were official in the U. S. Pharmacopoeia in 1898. It is considered toxic in overdose.

OTHER Above-ground parts cut off just above the root when flowers are coming into bloom dye wool green. Flower heads, with alum as a mordant, produce a yellow dye. Its strong scent helps keep away flies; adding Elder leaves makes an even better insect repellent. As the plant is obnoxious to insects, it makes a good biological control in home gardens.

TEASEL
Dipsacus sylvestris   Dipsaceae
NAMES This biennial European herb helped prepare wool for spinning; brushing with the stiff hairs of the dry flower head "teased" wool to straighten it. The Fuller's Teasel, *D. fullonum,* is better, also less likely to become a weed, but patches of the Common Teasel, *D. sylvestris,* found at sites of pioneer Oregon woolen mills indicate local use of this species. *Sylvestris,* of woods or forests, is not appropriate as this plant grows mostly in sunny open wasteland.

MEDICINE *Dipsacus,* from the Greek *dipsen,* to thirst, tells that the cup-like lower pair of joined leaves collects water. Water from these leaf-basins was used for cosmetics and as an eyewash. Herbalists from Dioscorides to Culpeper recommended Teasel; roots were considered diaphoretic, diuretic, bitter, and stomachic.

OTHER The relatively inconspicuous purply-pink blooms appear in July and August. Dried flower heads are popular for winter bouquets and assorted crafts; gloves are needed for working with these scratchy products.

Thimbleberry
Rubus parviflorus

THIMBLEBERRY
Rubus parviflorus   Rosaceae

NAMES Thimbleberry describes berry shape; *rubus,* Latin for red, fits the bright red berries; *parviflorus* says few flowers—or else it says small flowers, which these are not; (Latin, *parvus,* small). Unlike many Rose family plants, this shrub of moist shady areas has no thorns or prickles. The large white flowers appear in May, much later than those of its neighbor the Salmonberry.

FOOD In early spring Makah, Klallam, Swinomish, Samish, Upper Skagit, Chehalis, Snohomish, Quileute, Quinault, Squaxin, Coastal, Central and Southern Interior groups of British Columbia, ate Salmonberry sprouts. All ate the fresh berries. British Columbia groups considered these berries, difficult to pick in quantity, well suited to drying; most Washington Indians regarded them as too soft to dry well. Squaxin mixed fresh berries with Blackberries and dried them; Bella Coola often cooked them before drying. Nootka, who considered

Raspberries and Blackcaps better, mixed Thimbleberries with clams for a dried cake. Wildfood editors recommend these berries, fresh or cooked.

MATERIALS Deer graze on the plants, bear, birds and other wildlife eat the berries; bees and butterflies like the flowers. Cowlitz, Swinomish, Samish, Chehalis, and Quinault used the Maple-shaped leaves with Skunk Cabbage leaves to line baskets for preserving Elderberries, to separate different kinds of berries in one basket, and as a clean surface for drying berries; Okanagon used them to line steam-cooking pits. The large soft leaves were used for tissues and wiping. Shuswap and Carrier covered baskets of berries with leaves. Cowlitz used boiled bark as soap.

MEDICINE Makah boiled leaves for a solution to strengthen the blood and treat anemia. Vancouver Island Indians drank a decoction of roots and stems to arrest vomiting. Cowlitz applied powdered dry leaves to burns to avoid scarring; Upper Skagit treated swellings with grease mixed with ashes of burned leaves. No traditional or herbal medicine uses were found in sources consulted, though juice of the related Red Raspberry, *R. idaeus*, was in the National Formulary at times to carry and/or cover undesirable-tasting medicine.

OTHER Blackfoot of Alberta stained arrow quivers with Thimbleberries. Gardeners make many decisions in choosing plants to meet special needs; one source recommends Thimbleberry for a wild garden; another considers it a garden pest.

COMMON or BULL THISTLE; CANADA THISTLE
Cirsium vulgare; C. arvense, (Carduus spp.)   Compositae

NAMES *Cirsium* comes from the Greek *kirsos*, swollen veins, for which Thistles were considered a remedy. The Latin *Carduus*, formerly used for some *Cirsium* but now limited to Old World species of Thistles, is from the Celtic *ard*, a sharp point; this seems appropriate. Bull Thistle, *C. vulgare*, common, is an introduced European weed of roadsides and wasteland; the name Spear Thistle reflects this biennial's sharp prickers. Bull Thistle has a taproot.

Canada Thistle, *C. arvense*, of the fields, is a serious pest; this perennial is a native of Southeastern Europe and Asia, not Canada. Branching roots, source of the name Creeping Thistle, store food and help it travel. Male and female flowers are on different plants. Some colonies have only one kind of flower and are sterile; unfortunately this does not prevent spreading.

FOOD All Thistles are said to be edible if taken at the right time and properly prepared. Thistles are credited with saving the lives of several early explorers. Wildfood authors suggest roots boiled or roasted, or roots peeled, or the pithlike inner part of the root eaten raw or cooked. The species name of the native *C. edule* says that it is edible. Where it was available, native Americans ate peeled stems and young roots raw as salad or cooked as potherbs. A beverage tea was made from Thistle leaves. Willamette Valley Indians ate roots raw or roasted.

MEDICINE Traditional medicine formerly used roots or above-ground parts of Canada Thistle as diaphoretic, emetic, tonic, astringent, and emmenagogue. No herbal medicine uses were found. Lummi recognized Canada Thistle as adventitious, but boiled roots in salt water to drink at childbirth.

OTHER Thistles, many of which have purple flowers, attract birds and insects and provide bees with an excellent nectar source which makes good honey. Leaves of Thistles, crushed to destroy prickles, are good fodder for horses and cattle.

TIGER LILY
Lilium columbianum (L. parviflorum)   Liliaceae

NAMES *Lilium* came from *leirion*, classical Greek name for Lily; *columbianum*, of the Columbia river. Tiger probably came from the flower markings, though the bright red-orange flowers are spotted, not striped. This native Northwest perennial likes the sunshine of open woods and meadows, blooming in July.

FOOD Indians steamed and ate bulbs of several Lilies, including this one, raw or cooked, as we eat potatoes. As the bulbs are small, many are needed.

OTHER Although Tiger Lily will survive in cultivation, horticultural varieties are more available and more satisfactory.

TRILLIUM, WAKE-ROBIN
Trillium ovatum   Liliaceae

NAMES Plant parts in threes make the Latin *tri*, three, appropriate for Trillium. Wake-Robin calls attention to activity of birds at flowering time. *Ovatum* describes the broadly egg-shaped leaves. Many people feel spring has truly come when the white blooms of this perennial appear in March. Flowers are just above the whorl of three leaves on the unbranched stem rising from the root system, so picking removes the green leaves which manufacture the plant's food. The old wives' tale which says a Trillium, if picked, will not grow again for seven years confuses fact. If food remains in the rhizome and roots, the Trillium will send up a shoot the following year. This shoot will have one or more leaves, depending upon available stored food. One estimate says a Trillium grown from seed requires seven years to produce a bloom. It could also be several years before flowers reappear after picking, but the plant can bloom again when—or if—its strength is restored. Botanist Leonard Wiley reported a study of how picking affected flowering of Trilliums in his book Rare Wild Flowers.

FOOD Young leaves of Trillium have been used as cooked greens under the name "much hunger," a doubtful endorsement.

MEDICINE The name Birth Root and its variant Beth Root came from reported use of a similar species by Indians of eastern North America as a relaxant and pain reliever during childbirth. Lummi dropped juice of the bulb into sore eyes; Skagit soaked roots in water

Trillium
Trillium ovatum

for an eyewash. Quileute scraped the bulb on a boil to bring it to a head. Makah and Quileute had love charm uses. Stuhr says Indians used Trillium roots as an emetic and for female complaints.

Herbalists used a root poultice or leaves in lard as an external application for insect stings, and roots boiled in milk as a help in diarrhea. Traditional medicine included dried rhizome and roots of Trillium species in the National Formulary from 1917 to 1947 as treatment for hemorrhages and diarrhea; it was considered astringent, tonic, alterative, antiseptic, and emmenagogue.

OTHER A Quinault tradition said picking Trilliums brought rain. Trilliums grow well from seed and provide a cheery spring note in wild gardens; some nurseries stock plants. All *Trillium* species are natives of North America. Ants which enjoy an oil on the seed spread the plant by carrying away the seeds. Squirrels, chipmunks and other small animals also harvest the seeds.

BLACK TWIN-BERRY, BEARBERRY HONEYSUCKLE
Lonicera involucrata   Caprifoliaceae

NAMES Twin-berry refers to the paired fruit. Linnaeus named *Lonicera* for Adam Lónitzer, a German physician and naturalist, 1529-1586; *involucrata* says a circle of bracts or small leaves is below the flower cluster.

FOOD Wildfood editors report that these berries are edible, raw or cooked, good for pies, preserves, wine, or dried. They do not taste good to me, and I have not found any jam or jelly at wildfood outlets. Oregon Indians reportedly ate these, fresh or dried; Washington state and British Columbia groups apparently did not eat them; some agreed that bears liked them, confirming the name Bearberry Honeysuckle. Hitchcock says these are somewhat poisonous, but so bitter and nauseous that there is little danger of eating enough to cause trouble. Dennis says native *Lonicera* species have not been reported toxic, but that other species of *Lonicera* have poisoned those who ate them.

MEDICINE Quinault and Makah chewed leaves during pregnancy. Quileute chewed leaves as an emetic. Montana Flathead Salish ate berries as a laxative.

OTHER This native grows well at the Oregon coast; yellowish twin blooms appear in May, but the blue-black berries in deep-red bracts make the shrub unusual. Twin-berry is recommended for wild gardens; it is attractive to wildlife.

TWINFLOWER
Linnaea borealis   Caprifoliaceae

NAMES Flowers are twinned, a characteristic of the Honeysuckle family. This shrub was a favorite of Linnaeus, whose teacher, Fredrik Gronovius, had given it the name *Linnaea* in 1737.

MEDICINE Snohomish treated colds with a leaf decoction. An 1881 reference for botanical drugs noted use as a bitter, astringent, and antirheumatic; other herbal and traditional sources consulted did not mention it.

OTHER A small shrub or woody evergreen vine, Twinflower is native through much of the northern *(borealis)* hemisphere. Often found higher, it sometimes grows in shady areas at lower elevations, including Portland's Forest Park. Shiny dark green leaves, paired pink bell flowers, and sweet scent make this a desirable garden groundcover. It likes partial shade, and is fairly easy to propagate.

TWISTED STALK
Streptopus amplexifolius, S. roseus   Liliaceae

NAMES Both Large Twisted Stalk, *S. amplexifolius*, and Small, *S. roseus*, have unusual bends in the peduncle, an unexpected kink between stem and blossom. *Streptopus*, Greek for twisted, *pous*, foot or stem, for the zigzag stalks, *amplexi*, clasping, *folius*, leaf, as the leaves have no stems; *roseus*, for the rose-colored flowers. Leafy stalks of these two perennial natives are similar, but Small Twisted Stalk is perhaps half the size of the Large, and the rosy pink flower would not be confused with the white bell of the Large Twisted Stalk.

FOOD Berries, raw or cooked, can be eaten in limited quantities. Moderate consumption is logical if you know the common name Scootberry. Young stems have been eaten as salad greens. Various sources say Indians ate these berries; most British Columbia peoples considered them poisonous, and Gunther did not mention food uses by Western Washington groups. They are not found in great quantities, so may have had little use.

MEDICINE Makah women chewed and swallowed roots of Twisted Stalk to produce labor. Cowlitz treated tuberculosis with a leaf infusion.

OTHER These interesting Lilies, blooming in late April in moist coniferous woods, are suggested for wild gardens.

VANILLA LEAF, SWEET-AFTER-DEATH
Achlys triphylla   Berberidaceae

NAMES *Achlys*, Greek for mist, describes the white fluff of petal-less flowers topping the wiry, leafless bloom stems in April. *Triphylla* says the terminal whorl has three leaves. This genus has only two species; the other grows in Japan.

MATERIALS Pioneer ladies put these in their linen closets for the vanilla-like fragrance released by wilting leaves, reported by the names Vanilla Leaf and Sweet-after-Death. A vanilla substitute, impractical because of high production cost, has been extracted from leaves of this perennial.

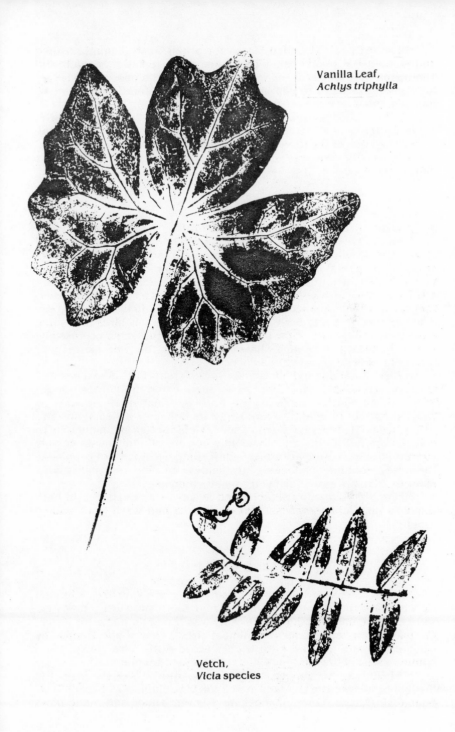

Vanilla Leaf,
Achlys triphylla

Vetch,
Vicia species

MEDICINE Skagit boiled leaves for a hair wash; Lummi mashed and soaked the plant, using the water as an emetic. Saanich and Thompson used leaves as a mosquito repellent. Thompson used leaves for washing bedding to eliminate bedbugs and other insect pests. A bunch of dried leaves is recommended for repelling flies.

OTHER Other common names are May-Leaf and Deerfoot. The name Butterfly Leaf describes the shape left after folding back the unmatched one of the three leaflets. Vanilla Leaf grows well from seed and spreads from underground stems; this plant of shady woodlands is nice in a wild garden.

VETCH
Vicia americana, V. sativa (V. angustifolia)   Leguminosae
NAMES *Vincire,* Latin, to bind together, refers to tendrils, typical of Vetches, which support the vines and help them climb. *Americana* says this perennial is a native; *sativa,* cultivated, infers the Common Vetch or Tare, a European perennial, was sown for hay. Vetches are attractive vines with several leaflets, (8 to 16), and a typical Pea flower; they usually have smaller flowers and leaves than do their cousins the Peas. Bumblebees and honeybees enjoy their nectar, birds and small animals eat the seeds. Many species have been cultivated extensively; some, including *V. sativa,* have escaped to become permanent residents. Several grow in our area.

FOOD American Indians ate seeds and stems of *V. americana* and various Vetch species; seeds have been eaten in Europe. South Kwakiutl ate roasted seeds of the Northwest native *V. gigantea* as snacks, but not in quantity. Kirk suggests boiled or baked stems and young seeds. Dennis says seeds of some *Vicia* species are dangerous to eat, though not all are equally deadly nor are all individuals equally susceptible. Some were introduced for feeding livestock, not people. As they may contain cyanogenic glycosides, caution and meticulous identification are essential if you consider food use.

MEDICINE Squaxin used crushed leaves of *V. americana* in bath water to take away soreness. Makah made a hair wash from soaked roots.

VINE MAPLE
Acer circinatum   Aceraceae
NAMES Stems of this shrub or small tree are vine-like. A branch tip which touches the ground may sprout roots and start a new tree, source of the name Walking Tree. *Acer* was the Latin name for Maple. *Circinatum,* circle, describes the almost-circular leaf shape. Handsome wine-red flowers hide beneath the leaves in April. Vine Maple, more common west of the Cascades, has canoe-shaped samaras.

MATERIALS Northwest Indians made utility baskets from the flexible branches; the Quinault name for Vine Maple was Basket Tree. Squamish, Katzie, Lower Lillooet, and Lower Thompson made bows

Vine Maple
Acer circinatum

from straight branches, arrows, implement handles, and dipnet frames from the strong supple wood. Quinault made fish traps, Skagit, salmon tongs. It provided fuel and digging sticks. Woodsmen like Vine Maple for pothooks and cooking sticks because it does not burn well.

OTHER Deer browse on leaves and twigs; squirrels and chipmunks eat the seeds. Passing through a group of these sprangling shrubs is such a challenge that French-Canadian trappers called it "devil wood." Quinault mixed Vine Maple charcoal with oil to make black paint. Vine Maple, valued for landscaping, is effective in groups; its brilliant fall colors are especially attractive.

EVERGREEN, PURPLE, WOOD VIOLETS
Viola sempervirens, V. adunca, V. glabella   Violaceae

NAMES *Viola* is an old Latin name for a similar species. Of the many in our area, three native perennials are relatively common. *Viola sempervirens*, Latin for always green, identifies the leaves, as does Evergreen Violet. Another name, Trailing Yellow Violet, says it spreads by runners. Shiny green leaves make *V. sempervirens* a good ornamental. Its flowers from March in moist woods west of the Cascades grow on short unbranched stems; the lower petals have purple lines.

The Purple Violet, *V. adunca*, blooms in woods or open spaces about March. *Adunca*, hooked, pictures the spur at the base of the lower petal; it is longer than similar spurs on other Violets. This blue to deep violet flower grows through much of western North America and is desirable for dry open woods or partly shady rockeries.

Many names for *V. glabella*, smooth in Latin, reflect how commonly its large clusters appear. Smooth Woodland Violet, Yellow Wood Violet, Pioneer Violet, and Stream Violet are all informative. Its leaves reach up toward light and air in somewhat open spaces, and its flowers are above the ground on branched leafy stalks from March to May and occasionally later. *V. glabella* is rather aggressive for a native wild garden. Because this is so widely available, it may be used more freely than other species without fear of destroying the plant. Do remember to ask permission before picking wild plants anywhere. See name comments,

171

FOOD Virgil and Homer mentioned food use of Violets. Sugared and preserved flowers make attractive garnishes; flowers and leaves of many *Viola* species can be used for seasoning or included in a spring salad; later in the season they may be tough. Leaves have been cooked for spring greens, dried leaves used for a hot beverage. Flowers contribute vitamins A and C as well as color and flavor in salads and desserts. Northwest Indians ate flowers raw, leaves cooked. A Violet, Pigweed, (*Chenopodium album*), and a Dock (*Rumex acetosa*), were among plants identified in stomach contents of the Tollund man, buried more than 2,000 years ago in a Danish peat bog.

MEDICINE Makah women ate *V. adunca* roots and leaves during labor; Klallam laid mashed flowers on the chest to treat pain. Herbalists have used leaves and flowers of various Violets medicinally, though Northwest species are not named in sources consulted. Roots of *V. pedata*, native to the eastern United States, were in the U. S. Pharmacopoeia in 1873 for use as emetic and cathartic. Leaves of the common European Pansy, *V. tricolor*, were in the U. S. Pharmacopoeia in 1883 as a mild laxative, emollient, and for pectoral complaints.

OTHER The broad tip of the lower petal of Violet flowers, marked with stripes pointing the way to nectar and pollen, is an excellent landing platform for pollinating insects. Violets provide a classic example of a self-pollinating technique, cleistogamous flowers which set viable seeds but usually do not open. (Greek *kleistos*, closed, *gamos*, marriage.)

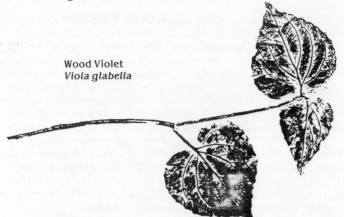

Wood Violet
Viola glabella

PACIFIC or SLENDER-STEMMED WATERLEAF
Hydrophyllum tenuipes   Hydrophyllaceae

NAMES Waterleaf translates the Greek *hydro*, water, and *phyllon*, leaf. Water caught and saved in the leaf cavity is one suggested name source; another is the somewhat splotchy leaves, which may appear to be water spotted. *Tenuipes* means slender or thin-footed. This species grows up to three feet; flowering stalks rise above the leaves. The

172

Pacific Waterleaf
Hydrophyllum tenuipes

Shady Waterleaf
Phacelia nemoralis

petal-less May flowers, greenish or pale white to blue, have many conspicuous stamens. The root system holds water so efficiently that Waterleaf plants often stay green after others have dried up; this has also been suggested as a name source. The plant dies back after seeds have matured. The frequency of chewed leaves suggest this large-leaved woodland native appeals especially to insects.

FOOD   Cowlitz ate roots of *H. tenuipes,* which grows mostly west of the Cascades. Thompson and Southern Shuswap of Interior British Columbia formerly cooked and ate roots of *H. capitatum,* Ball-head Waterleaf, widespread east of the Cascades. Sturtevant said native Americans of eastern North America ate spring shoots of *Hydrophyllum* species of their areas, and roots in times of food scarcity. Young shoots are suggested for salad, and leaves and roots cooked. The hairy leaves hardly seem an appealing food! Livestock eat leaves and roots, and it is often heavily grazed by wild game.

OTHER   Pacific Waterleaf causes contact dermatitis for some people. *H. tenuipes* is somewhat rank and aggressive for home gardens. *H. capitatum* is more suitable.

SHADY WATERLEAF, WOODLAND or SHADE PHACELIA
Phacelia nemoralis   Hydrophyllaceae

NAMES   The common names of this native perennial combine family and species with the plant's habitat, Greek *nemoralis,* of groves or woods. *Phacelia,* from the Greek *phakelos,* a fascicle or cluster, describes the crowded flower stems typical of some species. The flower spike is called a scorpioid raceme, as the stalks curl up like a scorpion's tail, unrolling as the plant develops. Flowers are white or pale bluish.

OTHER   Shade Phacelia grows in open shady woods, dry woods and plains, mostly west of the Cascades. Its bristly feel suits a plant which may cause irritant or contact dermatitis.

WATER PARSLEY
Oenanthe sarmentosa   Umbelliferae

NAMES   Water Parsley grows in wet areas west of the Cascades; somewhat Parsley-like leaves gave this native perennial its common name. *Oenanthe,* from the Greek *oinos,* wine, and *anthos,* flower, says some species were used in wine; *sarmentosa,* bearing runners, as the plant sends down roots at plant nodes.

FOOD   Lower Chinook, Snuqualmie, Cowlitz, Skokomish, and Cowlitz ate the tender young stems; Sturtevant said Oregon Indians ate boiled roots like potatoes. Current wildfood editors do not mention food use.

MEDICINE   Quinault formerly put warmed Water Parsley leaves on sore limbs. Makah used the pounded root as a laxative. Hitchcock says the root is a potent laxative, though other sources consulted neither suggest this use or warn against eating it. Herbal medicine formerly

174

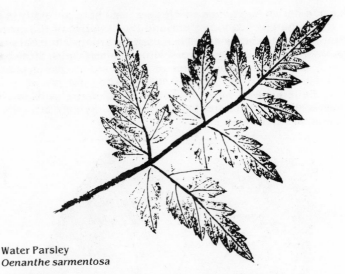

Water Parsley
Oenanthe sarmentosa

used other *Oenanthe* species for external applications; poisonous qualities were noted.

OTHER While Water Parsley, *O. sarmentosa*, is not listed as poisonous, its close resemblance to other Parsley family species, many poisonous, makes caution appropriate.

WESTERN WAHOO, BURNING BUSH
Euonymus occidentalis   Celastraceae

NAMES The Creek and Sioux name for the eastern shrub was Wahoo. *Euonymus*, a combination of the Greek *eu*, well or good, and *onoma*, name, inferring well known, may have been appropriate for the European species, but Western Wahoo seems more a curiosity than a common plant. Brightly-colored sepals and seed capsules gave the name Burning Bush. *Occidentalis*, of the west, is Latin; this species grows west of the Cascades. After it was included in an Endangered Species list, so many were located that it was removed from the list!

MEDICINE Herbalists used bark of large stems and roots of *E. occidentalis* for gall bladder and large intestine problems, as a cardiac stimulant, and a diuretic; bark, root, and berries of *E. atropurpureus* and *E. europaeus* also had herbal use. Large quantities are potentially toxic. Traditional medicine included dried root bark of the Wahoo, *E. atropurpureus*, in the U. S. Pharmacopoeia from 1863 to 1916 and the National Formulary from 1916 to 1947, as a gentle purgative, and for a mild effect on the heart similar to Digitalis.

OTHER *E. occidentalis* has been suggested for an ornamental. The flowers are attractive in June and the bright seedpods in late August and September.

Three non-native *Euonymus* species are used horticulturally in our area: *E. atropurpureus*, Wahoo or Burning Bush, native of the eastern United States; (young shoots have been used for artists' charcoal which is especially smooth and easy to work with); *E. americanus*, Strawberry Bush, and *E. europaeus*, Spindle Tree or Prickwood, native of Europe. Prickwood came from use of the strong wood for skewers and toothpicks. Eating leaves, bark, or fruit of these three shrubs or vines causes severe diarrhea, apparently not fatal. Children are attracted to the fruit, animals to the leaves.

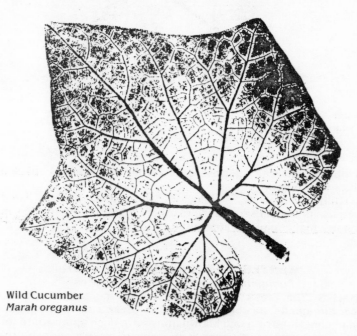

Wild Cucumber
Marah oreganus

WILD CUCUMBER, BIGROOT, OLD MAN IN THE GROUND

Marah oreganus (Echinocystis oregana)   Cucurbitaceae

NAMES   Bigroot, Manroot, and Old Man in the Ground picture size of the water-storing roots. *Marah*, Hebrew for bitter root, describes the very bitter root juices; *oreganus*, of Oregon. Wild Cucumber is related to melons and squashes, also trailing or climbing vines; like them, it has separate pollen-bearing and seed-bearing flowers; both grow on the same plant. This perennial vine, blooming from May, mostly west of the Cascades, is sometimes a nuisance in cultivated fields. Specimens collected by Scouler, Douglas, and Tolmie are part of early Northwest botanic records.

FOOD   In Wild Flowers of the Pacific Northwest, Clark says Indians may have eaten fruit of *M. oreganus*, as plants were found at sites of former Indian settlements in California and Oregon.

176

MEDICINE Lower Chinook used the fruit as a poultice; Squaxin mashed the upper stalk in water in which they dipped aching hands; Chehalis mixed ashes of the root with bear grease and applied this to scrofula sores. Hitchcock says about six *Marah* species are native to western United States and adjacent Canada and Mexico, but lists only *M. oreganus* for our area. Sweet says California Indians used oil from roasted seeds of a Wild Cucumber to promote hair growth, used roasted seeds for kidney treatment, and put crushed green roots in streams to stupefy fish.

OTHER Current authors consulted mention neither food or medicinal uses. Dennis, Kingsbury, and Muenscher, poisonous plant sources, included no *Marah* or *Echinocystis* species. Another source listed as poisonous seeds and root juice of *Marah fabacea*, one of five species which grow in California, and the 1950 Dispensatory said roots of *M. fabacea* were emetic and cathartic. Gray digger squirrels are reported to enjoy the cucumber-like fruits. They surely need them more than we do!

WESTERN WILD GINGER
Asarum caudatum   Aristolochiaceae

NAMES The ginger-like flavor of this May-blooming perennial gave the common name. The genus was named for a plant which wilts readily, so *Asarum*, from Greek *a*, not, plus *seiro*, to bind, says it is not suitable for garlands. Dark wine-colored sepals of the petal-less flowers of this species stretch out to a long thin thread, source of *caudatum*, with a tail. The eastern species, *A. canadense*, does not have these "tails."

FOOD Indians used roots for seasoning. Wildfood editors suggest using candied or dried roots instead of commercial ginger.

MATERIALS Thompson and Okanagan mixed Wild Ginger with Sphagnum moss for infant bedding valued for its pleasant scent. A volatile oil from Wild Ginger has been used in perfume.

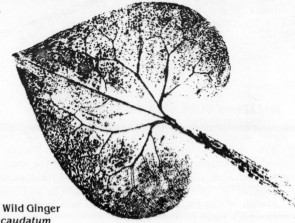

Western Wild Ginger
Asarum caudatum

MEDICINE Skagit treated tuberculosis with dried leaves; Upper Skagit boiled Wild Ginger for a tonic drink. Stuhr said female complaints were treated with a plant infusion. Traditional and herbal medicine used dried roots and rhizomes of *A. canadense* as stimulant, carminative, diuretic, stomachic, tonic, and diaphoretic. It was in the U. S. Pharmacopoeia from 1820 to 1873 and in the National Formulary from 1916 to 1947. Two antibiotic substances have been isolated from this drug. Leaves of a European species had other medicinal uses. Indians of eastern North America used roots of *A. canadense* as emmenagogue, stomachic, vulnerary, for throat trouble, abortive and contraceptive for temporary sterility, also in a powder used to prevent tooth decay.

OTHER Wild Ginger, available at some nurseries, is suggested as a dark green carpet in a moist shady garden. It has been on the Protected Plants List. A relative, *A. arifolium*, has been used in flavoring soft drinks.

Asarum is a relatively primitive and ancient genus; it appears in the fossil record before flies, bees, and butterflies; beetles were among the earliest insects. The three-sided flowers are beneath the leaves and low to the ground, a helpful location for creeping insects, long credited with providing pollination. Studies reported in the February, 1983 bulletin of the Native Plant Society of Oregon suggest these plants depend on cleistogamous, (self-pollinated), flowers for reproduction.

Wild Lily of the Valley
Maianthemum dilatatum

WILD or FALSE LILY OF THE VALLEY
Maianthemum dilatatum   Liliaceae

NAMES False or Wild Lily of the Valley says leaves of this native perennial are similar to those of the garden favorite. *Maios*, for the blooming period, plus *anthemon*, flower, gave the genus name; *dilatatum*, spread out, is an obscure reference. The heart-shaped leaves have a drip tip, useful in a moisture-laden location; it grows west of the Cascades. *M. dilatatum* has flower parts in twos and fours; Lily family plants more commonly have parts in threes.

FOOD Quileute, Lummi, Squamish, Kwakiutl, Bella Coola and Haida formerly ate the red berries, sometimes mixed with other berries in dried cakes stored for winter, but did not regard them highly. Wild Lily of the Valley was not common enough to be an important food source. Current wildfood writers do not consider these desirable for food; one listed the berries as poisonous.

MEDICINE Quinault soaked the pounded root of *M. dilatatum* in water to make an eyewash; Makah swallowed juice of pounded, chewed roots to correct sterility. Traditional and herbal medicine sources did not list this or other *Maianthemum* species, nor did Dennis, Muenscher or Kingsbury, poisonous plant sources consulted. Leaves, roots, fruit and flowers of the true Lily of the Valley, *Convallaria majalis*, are toxic. This Eurasian native contains cardioactive glycosides similar to those of Digitalis, but less toxic; these cause irregular heartbeats and stomach upsets. *C. majalis* is a source of contact dermatitis for some.

OTHER In woody settings where an aggressive ground cover could be helpful, the shiny-leaved perennial Wild Lily of the Valley is attractive. It spreads by underground stems and will grow in open shaded woods.

WILLOWS
Salix species   Salicaceae

NAMES *Salix*, from the Celtic *sal*, near, and *lis*, water, recognizes that Willows flourish in wet or poorly-drained soil. It is the classical Latin name for Willow. Male and female flowers (staminate and pistillate catkins), grow on separate plants. Willows are difficult to identify, as they interbreed extensively. We recognize them by pussywillows in the spring, often notice them near stream beds. Among species identified at Tryon Creek Park and the Camassia Wilderness near Portland are *S. lasiandra*, Long-leaf or Pacific Willow; *S. piperi*, Piper's Willow; and *S. scouleriana*, Scouler Willow. No western Willow species produces commercial timber, but Willow wood is good for campfires; many in our area are shrubs rather than trees. Kozloff has helpful comments about Willows.

FOOD Kirk suggests inner bark of Willows as an emergency food, less bitter if dried and ground into flour. Quileute used a few leaves in a cooking basket to flavor food. Deer and livestock browse on Willow; rabbit, mice, beaver, grouse and other small animals eat the bark. Willows provide substantial food supplies to caterpillars of many butterflies and moths.

MATERIALS Quileute, Klallam and Chehalis used bark for string, the twigs or osiers for basketry materials. Quinault made a heavier line for tumplines and slings, also harpoon lines for seal hunters, from Willow bark. Small trees were chosen for fish weirs as they took root.

MEDICINE Makah washed hair in water in which Willow roots had been soaked; Klallam boiled bark to treat sore throat and tuberculosis. Indians, not only in the Northwest, have long used Willows as a medicinal source, predating aspirin. Herbal and traditional medicine recommended Willow bark to treat headaches, fevers, as a tonic, and

179

Wood Fern
Dryopteris austriaca

antiseptic. Active ingredients are salicin and tannin. Aspirin is acetylsalicylic acid. Willow bark was in the U. S. Pharmacopoeia in 1880 as a bitter, an astringent, and febrifuge. Most species are said to have similar medical properties; no attempt has been made to make a complete listing of uses.

OTHER   In the Middle Ages, when dwarves were highly desirable, it was a folk belief that striking a child with Willow wands would stunt growth. Willows are useful for erosion control.

WINDFLOWER, THREE-LEAVED ANEMONE
Anemone deltoidea   Ranunculaceae

NAMES   Pliny said Anemones were sent by Anemos, god of the winds, as heralds of his coming; folklore says these spring flowers open only when the wind is blowing! Seeds are spread by the wind. The Greek capital letter *delta* is triangular; Greek *eidos* means form, so the three-part triangular leaf provides both common and species names. This perennial Wood Anemone blooms in May in shady moist woods west of the Cascades; its petal-less flowers have white sepals.

MEDICINE   Cowlitz treated tuberculosis with a windflower tea; Lillooet and Kootenay used an Anemone as a counterirritant. Indians of eastern North America used roots and leaves of Anemones for local irritants. Traditional and herbal medicine have used *Anemone* species; *A. patens* was in the U. S. Pharmacopoeia from 1882 to 1905 and the National Formulary from 1916 to 1947 as diuretic, expectorant, and emmenagogue. A powerful local irritant drug has been obtained from the dried herb of some Anemones; fresh leaves were preferred for making a tincture.

OTHER   Garden Anemones, highly regarded by Gerard and Culpeper, were formerly used in herbal medicine. Dennis reports three non-native *Anemone* species grown in Oregon gardens are dangerous if eaten, because of the amounts of protoanemonin, an acrid irritant substance; Oregon species apparently have not been tested.

SPREADING WOOD FERN, SHIELD FERN
Dryopteris austriaca (D. spinulosa, D. dilatata)   Polypodiaceae

NAMES   Wood Fern likes acid soil in moist shady woods and often grows on rotting wood. *Dryopteris,* Greek *drys,* Oak, *pteris,* fern, says some species grow near Oaks; *austriaca:* from Austria. Fronds of this lacy fern are basically triangular and a somewhat darker green than Lady Fern. Wood Fern leaves or fronds are tripinnate, as the small leaflets are divided and then divided again. Leaves may persist through the winter without withering. See Lady Fern.

FOOD   Haida, Nootka, Cowlitz, and Puget Sound tribes dug and ate rhizomes in the fall and winter. Niska and Lower Lillooet of Interior British Columbia, where this fern is less common, ate the roasted and peeled rootstocks in the spring. Current wildfood authors do not regard this as one of the more desirable food ferns.

MEDICINE   Snohomish washed hair in water in which leaves had been soaked. Klallam treated cuts with the pulp of powdered roots.

OTHER   Kwakiutl and Coast Salish used roots of Spreading Wood Fern as tinder in a slow match which would smoulder for several days in a clam or mussel shell. This fern is attractive for partially shady gardens.

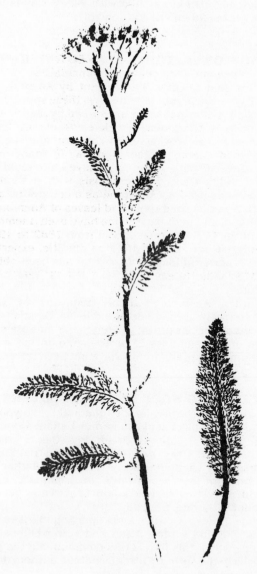

Yarrow
Achillea millefolium

182

YARROW, MILFOIL
Achillea millefolium   Compositae

NAMES Yarrow is from the Anglo-Saxon name for this plant, *gearwe.* The genus name reflects the belief that Achilles treated wounds of soildier with this herb during the siege of Troy. Soldiers' Woundwort and Nosebleed identify it as a vulnerary. *Millefolium,* thousand leaves, and the abbreviation Milfoil, refer to the finely-cut leaves. Yarrow grows in dry open woods and brushy disturbed areas, blooming from April to October.

FOOD Europeans ate Yarrow in salads during the 17th century; current wildfood editors do not suggest it, though some recommend dried or fresh leaves and seeds for a beverage tea. Yarrow was used as a substitute for Hops in the manufacture of beer in Sweden in the 18th century. Cattle avoid Yarrow, but if forced to eat it by lack of other food, produce milk with an unpleasant flavor.

MEDICINE This European perennial reached eastern North America by the 1600s, and no doubt soon came to the Pacific Northwest. Klallam and Squaxin used chewed or mashed leaves as a poultice for sores; Quileute prepared a poultice for skin rashes from boiled leaves. Thompson roasted leaves and stems and dusted the dried powder on sores. Cowlitz soaked leaves and used the water for washing hair. Makah and Klallam used Yarrow during childbirth; Quinault boiled roots for a tuberculosis treatment, or an eyewash; Cowlitz and Squaxin treated stomach trouble with the same decoction. Chehalis, Skagit and Snohomish boiled leaves for a diarrhea treatment.

Spanish Californians stopped bleeding of recent wounds with fresh plants and steeped leaves in hot water to treat cuts and bruises. European herbalists have recommended the dried plant, or oil from it, as a hair treatment. Herbal medicine has used a tea from the dried upper leaves and stems for a stomach tonic, and leaves as a poultice for rashes. Fresh leaves were chewed to cure toothaches. Large doses are considered poisonous. It causes contact dermatitis for some people. Dried flowering tops were in the U. S. Pharmacopoeia from 1863 to 1882 as stimulant, tonic, and emmenagogue. *A. millefolium* is more common west of the Cascades. A native species, *A. lanulosa,* east of the Cascades, had similar uses.

OTHER Haida strung clams on Yarrow stems for drying. Okanagan made a smudge of leaves on hot coals to repel mosquitoes. Dried flowers are popular for floral arrangements.

PACIFIC or WESTERN YEW
Taxus brevifolia   Taxaceae

NAMES The Greek *taxos,* Yew, is from *toxon,* bow, for which the wood was used in antiquity. *Brevifolia,* short leaves, describes the sharp-pointed flat needles, the darkest green of western evergreen needles. Yew, seldom abundant, grows well in the shade and can survive with minimum light. The seeds do not have a fibrous cover or cone; the somewhat pulpy, berry-like body is called an aril; (medieval Latin *arillus,* a wrapper for a seed).

MATERIALS Some Northwest Indians extended their smoking tobacco with Yew. Many valued Yew for archery bows;.others used the strong, durable wood in shafts for whale harpoons, as wedges for splitting logs and for digging sticks, frames of drums, spring poles of deer traps, paddles, and handles of tools. Mat-sewing needles, awls, fish hooks, knives, dishes, spoons, spears and spear points, boxes, dowels, pegs, fire tongs, and combs were also made of Yew.

MEDICINE Swinomish youth rubbed themselves with Yew sticks to gain strength; Chehalis soaked leaves in water used to bathe babies and old people. Klallam drank a leaf decoction for internal pain. Cowlitz ground moistened leaves and applied the pulp to wounds. Quinault chewed leaves and spit them on wounds and stings to promote healing, boiled dried bark and used the liquid as lung medicine. Haida and Fraser River Lillooet ate limited quantities of the pulpy fruits, the least harmful part.

OTHER Okanagan ground dry Yew and mixed it with fish oil to make a red paint. Dried Yew needles were sometimes mixed with smoking tobacco to extend supplies.

American Yew or Ground Hemlock, *T. canadensis;* English Yew, *T. baccata;* and Japanese Yew, *T. cuspidata,* as well as the native *T. breifolia,* are frequently used for garden ornamentals. Foliage, seeds, and bark of all the non-native species contain the alkaloid taxine and are usually toxic, so eating is not recommended. Toxicity varies with the species, the season, and the location. The native species has apparently not been tested for poisons, but it seems reasonable to avoid eating any parts of Yews. Muenscher reports that illness and deaths among cattle have been attributed to eating foliage and twigs of *T. brevifolia* in large quantities.

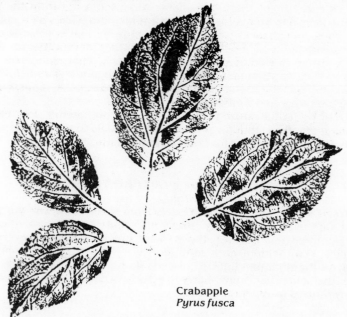

Crabapple
Pyrus fusca

184

Weave together the threads of your many interests. Do you enjoy a garden? How many plants you lovingly cultivate grow wild somewhere? Where did they live as natives? Have horticulturalists altered, (improved?), them? Can they still reproduce by seed?

If you are a hunter, do you know the favored foods of wild animals you seek? Are these native birds, animals or fish, or have they been introduced? Where was their original home? How fragile is their present location and how can you help protect it? Is it endangered by "civilization"?

Is needlework one of your pleasures? Some leaf shapes are incredibly beautiful. Even if you can't draw you can develop plant patterns. Try tracing a leaf; draw around it, or cover it with thin paper and make a rubbing. Transfer this to cloth and highlight it with embroidery. Look at wildflower colors; are these traditional for needlework? Are leaf sizes too large? With photocopy reduction you can fit several leaves on one needle painting.

At the turn of the century it was stylish to collect plants as botanists do type specimens; for years botany classes collected in hectic unconcern for wild plants. An ecologically sound way to collect plants is with photographs. Wind, light, and changing backgrounds make plant photography a pleasure for the beginner and a challenge for the experienced photographer.

Have you noticed familiar plants on stamps? How it impressed me to find a Dandelion on a stamp from Iceland! The United States recently issued stamps showing state birds with flowers. A Canadian series of province flowers in the 1960s included a Trillium. Switzerland has featured both wild and cultivated flowers. Many countries have pictured Foxglove, usually with medicinal plants. And the Orchids! We have few wild Orchids in the Pacific Northwest, but can enjoy them on stamps. Some flowers on stamps are pictorial, some stylized, some show both fruit and flower parts. Look for them.

Are you a calligrapher? Why not look at early herbals, advertising, live flowers and/or stylized plants in art and include a few in your work? An architect? Plants have long been appreciated for their form; some traditional designs incorporate plants, as Acanthus leaves in early Greek work. A genealogist? How many flowers are parts of coats of arms? Do you enjoy cake decorating? Use wildflower models for variety.

Leaf prints are fun as a craft project and the fine detail helps identify plants. The illustrations were made with a roller or brayer used for woodblock prints. Also needed are a smooth surface, block printing ink, paper, and leaves. Some leaves are fragile; Maple leaves are strong and make attractive prints. Use whatever paper is handy. Typing paper is satisfactory, but art paper can be more elegant.   Place a sheet of clean paper on a newsprint cushion; (flexibility of a few newspapers underneath the printing surface improves the print.) Squeeze ink on the smooth surface and roll it to an even consistency with the brayer.

Place a leaf, top side down, on a pad of scrap paper and roll over the back with the inked roller. (Veins are more prominent on the back.) Move the leaf to the clean paper, (the printing surface), placing it ink side down, and cover with scrap paper. (I use old newspaper.) Roll over the scrap paper with a clean brayer, or rub the scrap paper with your hand. Remove the scrap paper, lift the leaf, and there is your print. Experiment! Variations are fun.

For another kind of print, cover a leaf, (again, back side up), with a not-too-thick paper, and rub with crayon. More than one color can easily be used. To make a crayon lump similar to those used for brass rubbings, pour crayons, melted over hot water, into cupcake pans.

Honeysuckle
Lonicera ciliosa

BACKGROUND BUILDERS

Families

People

Books

Glossary

Index of Plants

English Daisy,
Bellis perennis

Wild Lettuce
Lactuca biennis

FAMILIES

Plant families have more similar features than the next smaller classification unit or taxon, the genus. The International Code of Botanical Nomenclature, (Latin, *nomen*, name), says family names should come from a major genus and end with -aceae, but permits continued use of existing names. Efforts are being made to phase out nonconforming names in use before the code was established in 1867. The familiar and descriptive older names are given here with newer names in parentheses.

Botanists arrange families in a sequence intended to show evolutionary development. In no case are these alphabetical. Botanically correct listings have separate sections for Ferns and Fern Allies, (2 families here, Ferns and Horsetail), gymnosperms, (2 families here, Pines and Yew), and angiosperms. There are roughly 375 families of flowering plants worldwide; about 50 of these are represented in the text, many by only one species. These families have more species than those mentioned; there are many other families in our area. The list here is alphabetical, not phylogenetic.

Berberidaceae; Barberry family species grow mostly in the Northern hemisphere; several have horticultural value. Indians had food, medicinal, and materials uses for Oregon Grape, also called Holly Grape.

Caprifoliaceae; Honeysuckle family plants, usually shrubs or woody vines, grow mostly in the North Temperate zone or mountainous regions in the tropics. Flowers and fruits are often twinned. Elderberries were an important food source for native Indians.

Compositae, (Asteraceae), Composite, Sunflower, or Aster Family, (Latin, *com*, together, *ponere*, to put with), is usually considered the largest family of flowering plants; distributed worldwide, they are primarily herbs. Its placing in the Englerian order indicates advanced development. The flower heads, composed of many flowers borne close together, resemble single flowers. "Ray" flowers around the edge attract pollinating insects, and "disk" flowers in the center produce seeds. The central flowers usually have petals fused into a tube; outer flowers are irregular, contributing to the petal-like effect of the cooperative arrangement.

The many similar plants in this family are often difficult to differentiate, so tribes, a smaller classification division, are sometimes used to help separate different groups. This family, not noted for food or drugs, includes Endive, Artichoke, Lettuce, Sunflower, and Safflower, as well as the ubiquitous Dandelion. Among the many flowers characteristic of the family are Asters, Dahlias, Chrysanthemums, Zinnias, Marigolds, Gaillardia.

Cruciferae, (Brassicaceae), cross-bearers, (Latin, *crux, crucis*, cross): one of the largest families, all herbs, also distributed worldwide. The many small regular flowers are often cross shaped, with parts usually in fours. The unique seedpods are above the flowers. Cruciferae are important economically; they include ornamentals, vegetables—and noxious weeds. *Brassica*, a genus important for foods, includes Cabbage, Cauliflower, Broccoli, and Mustard. Some cultivated vegetables, one being Radish, have escaped and become aggressive weeds.

Cucurbitaceae, Cucumber or Gourd Family, (Latin, *cucurbita*, gourd), are vines, distributed worldwide in warmer regions, often climb by tendrils; the fruit is characteristically a pepo—a nonopening, many-seeded berry with a hard rind.

Cyperaceae, Sedges, are herbs, not jointed, with small, dry, one-seeded fruit, usually grow in wet places, are often triangular in cross section—have edges!

189

Ericaceae, Heath family, mostly trees, herbs or shrubs; those in our area are usually shrubs. (Latin *erice*, heath.) Rhododendron is well known, but the family has other ornamental flowers, shrubs, and groundcovers. Indians ate large quantities of Salal and Huckleberries.

Geranium family, **Geraniaceae,** primarily herbs, have 5 sepals, 5 petals, 10 stamens in two circles, and a pistil with 5 parts; the seed case is a conspicuous characteristic of the family. Greek *geranion*, crane.

Graminae, (Poaceae), Grass Family, (Latin, *gramen*, grass), are herbs, jointed, mostly circular in cross section, usually growing in dry places. This family, important economically, is a major food source. The small, dry, one-seeded fruit, a grain, does not open at maturity and the seed coat is fused with the ovary wall. (Latin, *granum*, grain). Grasses, most successful and widespread of herbs, grow over vast areas, stabilize the soil, provide staple foods for man and grazing animals. Wheat, rice, corn, oats, rye, barley, sugar cane and bamboo are grasses. The family also includes many undesirable weeds. The lightweight pollen of grasses contributes appreciably to hay fever problems; a single plant may produce millions of pollen grains.

Grossulariaceae, Currant or Gooseberry Family, are largely shrubs. Currants are said to be unarmed; Gooseberries have thorns—are armed. Berries are a good food source.

Iridaceae, Iris family, has perennial herbs with showy, sometimes irregular, flowers. Flower parts are in threes and the fruit always a many-seeded capsule. Iris are valuable garden plants.

Labiatae, (Lamiaceae), from the Latin for lip; plants of the Mint family, considered highly advanced in evolutionary development, are usually herbs or shrubs. Plants of this large family are aromatic, have square stems, opposite leaves, and two-lipped flowers, the upper lip usually two-lobed and smaller than the three-lobed lower one, a convenience for pollinating insects. The showy characteristic flowers cluster in axils of leaves or form spikes at the top of the stem; some are garden favorites. Economic importance of the Mint family is due to volatile oils from such plants as Peppermint, Sage, Thyme, Lavendar, and Spearmint.

Leguminosae, (Fabiaceae), came from the Latin *leger*, to gather, as the unique Pea family fruits, dry pods opening along two side seams, can be picked without cutting. One of the largest families of flowering plants, Peas are worldwide in distribution and include herbs, shrubs, vines, and trees. It is an important food source, providing Peas, Lentils, Clover, Alfalfa and other forage crops. There are also ornamentals, plants used for cut flowers, for shade, and tropical species which provide lumber. All, helped by bacteria associated with their roots, are able to take nitrogen from air in the soil and fix it in a form available for use. Leaves are usually pinnately compound; sometimes terminal leaflets are modified to tendrils. We all know the Peas and the typical Sweet Pea flower, shaped somewhat like a butterfly.

Plants of the Lily family, **Liliaceae,** provided the best supply and the most useful root food to Indians throughout North America. Lilies, largely perennial herbs, have parts in threes, showy flowers, mostly many-seeded fruit; principal veins of the leaves are parallel or nearly so; many grow from bulbs. Many Lilies are ornamentals—as Tulips, Hyacinths, Day Lilies—but Onions and Asparagus are also in this family. There are some drugs, some poisonous species.

Orchidaceae, all herbs, commonly grow in tropical areas, and are the largest family of monocots, having more than 20,000 species in more than 600 genera. (Greek *orchis*, testicle.) The typically irregular flowers are highly specialized for insect pollenization; look for their landing platform! As the tiny seeds store little food, they need especially suitable growing conditions for successful germination; many also require cooperation from associated fungi. Economic importance comes from the many ornamentals; Vanilla is produced from one of this family.

Polypodiaceae, Ferns, reproduce by spores; their life cycle includes two types of plants. The handsome plants we recognize are the sporophyte generation, which produces spores seen as dots on the back of fern leaves or fronds. Each species has a different spore pattern. The tremendous number of spores on each spore-producing frond reflects the limited chance of success for individual spores. A spore which falls to a favorable location develops an inconspicuous leaflife "prothallium," the gametophyte generation, which produces the sexual parts which unite to develop the fern we know. Water to bring together male and female parts is essential to the reproductive process. Ferns live mostly in moist shady habitats.

Terms applied especially to Ferns are: pinna, (pinnae, pl.), Latin, *pinna*, a feather, a leaflet or division of a compound leaf or frond; pinnule, (diminutive of *pinna*) the secondary leaflet of a fern, one part of a twice-pinnate leaf; and rachis, (Gr. *rachis*, a backbone), the stem of a fern frond from which pinnae arise.

Ranunculaceae, Crowfoot or Buttercup family, mostly herbs and a few woody vines, grow in north temperate and arctic regions; this family is tremendously successful. Botanists consider it one of the more primitive angiosperm families. Many contain poisonous alkaloids. Flowers are generally regular and leaves often much divided. Ornamental flowers include Delphinium, Anemone, Columbine, Clematis, Peony, and Buttercups.

Rosaceae—Rose family—include trees, shrubs, and herbs, and differ in many ways. Many are native to the northern hemisphere; many have compound leaves; many are popular with insects because they produce generous quantities of pollen. All wild Roses have flowers with five petals and stamens in a ring around the edge of a sunken cup. Rosaceae have varied fruit types: achenes, pomes, drupes, capsules, and follicles. Some have great economic importance: Apples, Cherries, Strawberries, Almonds. Roses and other family members—Hawthorn, Laurel Cherry, Mountain Ash—are extensively used for ornamentals. Horticulturalists use rootstocks of wild Roses, called species roses, for grafting the more fragile garden varieties; wild Roses are less susceptible to disease.

The Warm Springs Indian name for Rose means "mean old lady, she stick you!" Botanists say that plants with protective devices such as prickers or thorns are armed; armor varies with the species. Rose thorns are adaptations of the bark; they break away cleanly. Prickles are small sharp points or superficial outgrowths of the stem; Anglo-Saxon *prickele*, prickle.

Rubiaceae, Madder family, is one of the larger families of angiosperms. Most grow in the tropics, many are trees. Those in our area are herbs. Coffee belongs to this family, as do the Gardenia and Cinchona or Peruvian bark, source of quinine.

Salicaceae, Willow family, contains only two genera, *Salix*, and *Populus*, both represented in our area. There are many species in the tropics; all like moist growing conditions. The wind-pollinated flowers, male and female on different plants, appear in catkins in spring before the trees leaf out.

Saxifragaceae: saxifrage means rock breaker: Latin *saxum*, rock, and *frangere*, to crack. The name may come from herbal use of plants to treat "stones" of the urinary tract. I find more interesting an alternate source: many species grow in rock crevices where the growing roots exert a great deal of pressure. These herbs, widely distributed in the northern hemisphere, often have palmately-veined notched leaves and small flowers on tall stems or scapes.

Scrophulariaceae, Figwort or Snapdragon family, largely herbs, occasionally shrubs or vines, includes garden and greenhouse flowers: Penstemon, Calceolaria, Foxglove. Many species have had medicinal use.

Solanaceae, Potato or Nightshade family, is notable for food use, including Potato, Tomato, and Eggplant; Tobacco and Petunias are also in this family. The fruit is usually a berry. The family is known for poisons, including *Atropa belladonna*, Deadly Nightshade, used medicinally.

Umbelliferae, (Apiaceae), describes the arrangement of Parsley or Carrot family flowers. Pedicels (stems) of the many small flowers arise from about the same point and form a large flat or rounded cluster—an umbel. This is designed to attract pollinating insects, one of which is the Black Swallowtail Butterfly. Plants, mostly herbs, often have hollow stems, more or less divided leaves, and dry, one-seeded fruits. This family has many poisonous plants, but also food plants: Carrots, Parsnips, Parsley.

Aceraceae, Maple Family
Bigleaf Maple   *Acer macrophyllum* Pursh
Vine Maple   *A. circinatum* Pursh
Anacardiaceae, Sumac Family
Poison Ivy   *Rhus radicans* L.
Poison Oak   *R. diversiloba* T. & G.
Apocynaceae, Dogbane Family
Dogbane   *Apocynum androsaemifolium* L.
Aquifoliaceae, Holly Family
English Holly   *Ilex aquifolium* L.
Araceae, Arum or Calla Lily Family
Skunk Cabbage   *Lysichitum americanum* Hulten & St. John
Araliaceae, Ginseng Family
English Ivy   *Hedera helix* L.
Aristolochiaceae, Birthwort
Wild Ginger   *Asarum caudatum* Lindl.
Berberidaceae, Berberis Family
Duckfoot   *Vancouveria hexandra* (Hook.) Morr. & Dec.
CreepingOre.Grape   *Berberis repens* Pursh
Low Ore.Grape   *B. nervosa* Pursh
Tall Ore.Grape   *B. aquifolium* Pursh
Sweet-after-Death   *Achlys triphylla* (Smith) DC.
Betulaceae, Birch Family
Oregon or Red Alder   *Alnus rubra* Bong.
Western Hazel   *Corylus cornuta* Marsh.
Caprifoliaceae, Honeysuckle
Blue Elderberry   *Sambucus cerulea* Raf.
Red Elderberry   *S. racemosa* L.
Twinflower   *Linnaea borealis* L.
Orange Honeysuckle   *Lonicera ciliosa* (Pursh)DC.
Twin-berry   *Lonicera involucrata* (Rich) Banks
Snowberry   *Symphoricarpos albus* (L.) Blake
Caryophyllaceae, Pink Family
Bladder Campion   *Silene cucubalis* Wibel
Chickweed   *Stellaria media* L.

Celastraceae, Staff-tree Family
W. Wahoo   *Euonymus occidentalis* Nutt.
Chenopodiaceae, Goosefoot Family
Lamb's Quarters   *Chenopodium album* L.
Compositae (Asteraceae), Aster
Douglas Aster   *Aster subspicatus* Nees
Bull Thistle   *Cirsium vulgare* (Savi) Airy-Shaw
Common Burdock   *Arctium minus* (Hill) Bernh.
Great Burdock   *A. lappa* L.
Canada Thistle   *Cirsium arvense* (L.)var. *horridum* Wimm. & Grab.
Hairy Cats-ear   *Hypochoeris radicata* L.
Smooth Cats-ear   *H. glabra* L.
Chicory   *Cichorium intybus* L.
Coltsfoot   *Petasites frigidus* (L.)Fries
Daisy Fleabane   *Conyza canadensis* (L.) Cronq.
English Daisy   *Bellis perennis* L.
Dandelion   *Taraxacum officinale* Weber
Dogfennel   *Anthemis cotula* L.
Goldenrod   *Solidago canadensis* L.
West.Goldenrod   *S. occidentalis* (Nutt.)T.&G.
Common Groundsel   *Senecio vulgaris* L.
Hawkbit   *Leontodon autumnalis* L.
Hairy Hawkbit   *L. nudicaulis* (L.)Merat
Rough Hawksbeard   *Crepis setosa* Hall f.
Smooth Hawksbeard   *C. capillaris* (L.) Wallr.
Prickly Lettuce   *Lactuca serriola* L.
Wild Lettuce   *L. biennis* (Moench) Fern
Nipplewort   *Lapsana communis* L.
Oxeye Daisy   *Chrysanthemum leucanthemum* L.
Pathfinder   *Adenocaulon bicolor* Hook.
Pearly Everlasting   *Anaphalis margaritaceae* (L.)B.&H.
Purple Salsify   *Tragopogon porrifolius* L.
Yellow Salsify   *T. dubius* Scop.

Sow Thistle *Sonchus arvensis* L.
Prickly Sow Thistle *S. asper* (L.) Hill
Tansy Ragwort *Senecio jacobaea* L.
Tansy *Tanacetum vulgare* L.
Yarrow *Achillea millefolium* L.
Convolvulaceae, Morning Glory
Morning Glory *Convolvulus arvensis* L.
Cornaceae, Dogwood Family
Canada Dogwood *Cornus canadensis* L.
Creek Dogwood *C. stolonifera* Michx.
West. Fl. Dogwood *C. nuttallii* Aud.
Cruciferae (Brassicaceae), Mustard
W. Bitter Cress *Cardamine oligosperma* Nutt.
Wood Bitter Cress *C. angulata* Hook.
Garlic Mustard *Alliaria officinalis* Andrx.
Honesty *Lunaria annua* L.
Cucurbitaceae, Cucumber or Gourd
Wild Cucumber *Marah oreganus* (T.&G.) Howell
Cupressaceae, Cypress Family
Western Red Cedar *Thuja plicata* Donn.
Cyperaceae, Sedge Family
Carex species
Scirpus species
Dipsacaceae, Teasel Family
Common Teasel *Dipsacus sylvestris* Huds.
Equisetaceae, Horsetail Family
Common Horsetail *Equisetum arvense* L.
Giant Horsetail *E. telmateia* Ehrh.
Scouring Rush *E. hyemale* L.
Ericaceae, Heath Family
Evergreen Huck. *Vaccinium ovatum* Pursh
Red Huckleberry *V. parvifolium* Smith
Indian Pipe *Monotropa uniflora* L.
Madrona *Arbutus menziesii* Pursh
Salal *Gaultheria shallon* Pursh
Fagaceae, Beech Family
Garry Oak *Quercus garryana* Dougl.
Fumariaceae, Fumitory Family
West.Bleeding Heart *Dicentra formosa* (Andre.)Walpers.
Geraniaceae, Geranium Family
Filaree *Erodium cicutarium* (L.)L'Her.
Cut-leaf Ger. *Geranium dissectum* L.
Dovefoot Geranium *G. molle* L.
Gramineae (Poaceae), Grass Family
Grass species
Grossulariaceae, Currant
Red Fl.Currant *Ribes sanguineum* Pursh
Hydrangaceae, Hydrangea Family
Syringa *Philadelphus lewisii* Pursh
Hydrophyllaceae, Waterleaf Family
Grove Lover *Nemophila parviflora* Dougl.

Shady Waterleaf *Phacelia nemoralis* Greene
Sl. St. Waterleaf *Hydrophyllum tenuipes* L.
Hypericaceae, St.John's Wort Family
St. Johnswort *Hypericum perforatum* L.
Iridaceae, Iris Family
Oregon Iris *Iris tenax* Dougl.
Labiatae (Lamiaceae), Mint Family
Creeping Charlie *Glecoma hederacea* L.
Heal-all *Prunella vulgaris* L.
Lemon Balm *Melissa officinalis* L.
Red Dead-Nettle *Lamium purpureum* L.
Hedge Nettle *Stachys cooleyae* Heller
Leguminosae (Fabiaceae), Pea
Red Clover *Trifolium pratense* L.
Lupine *Lupinus* species
Scotch Broom *Cytisus scoparius* (L.)Link
White Clover *Trifolium repens* L.
Perennial Pea *Lathyrus latifolius* L.
American Vetch *Vicia americana* Muhl.
Common Vetch *V. sativa* L.
Liliaceae, Lily Family
Fairy Bell *Disporum hookeri* (Torr.) Britt.
Fairy Lantern *D. smithii* (Hooker)Piper
False Hellebore *Veratrum californicum* Durand
Green False Hellebore *V. viride* Ait.
False Sol.Seal *Smilacina racemosa* (L.)Desf.
Star-flrd. F.Sol.Seal *S.stellata* (L.)Desf.
Fawn Lily *Erythronium oregonum* Applegate
Wild Lily of Valley *Maianthemum dilatatum* (Wood)Nels. & MacBr.
Tiger Lily *Lilium columbianum* Hans.
Trillium *Trillium ovatum* Pursh
Lg.Twisted Stalk *Streptopus amplexifolius* (L.)DC.
Small Twisted Stalk *S. roseus* Michx.
Oleaceae, Olive or Ash Family
Oregon Ash *Fraxinus latifolia* Benth.
Onagraceae, Evening Primrose Family
Enchanter'sNightshade *Circaea alpina* L.
Eve.Primrose *Oenothera biennis* L.
Red-sepaled Eve.Prim. *O. erythrosepala* Borb.
Fireweed *Epilobium angustifolium* L.
Tall An.WillowHerb *E.paniculatum* Nutt.
Watson'sWillow Herb *E.watsonii* Barbey
Orchidaceae, Orchid Family
Bog Orchid *Habenaria dilatata* (Pursh)Hook.

Spotted Coral-root   *Corallorhiza maculata* Raf.

Lady'sSlipper   *Calypso bulbosa* (L.)Oaks

Phantom Orchid   *Eburophyton austiniae* (Gray)Heller

Rattlesnake Plantain   *Goodyera oblongifolia* Raf.

Oxalidaceae, Oxalis or Wood Sorrel
Oregon Oxalis   *Oxalis oregana* Nutt.
West.Yel.Wood Sorrel   *Oxalis suksdorfii* Trelease

Papaveraceae, Poppy Family
California Poppy   *Eschscholzia californica* Cham.

Pinaceae, Pine Family
Douglas Fir   *Pseudotsuga menziesii* (Mirbel)Franco
Grand Fir   *Abies grandis* (Dougl.)Forbes
West.Hemlock   *Tsuga heterophylla* (Raf.)Sarge.
Lodgepole Pine   *Pinus contorta* Dougl.
Ponderosa Pine   *P. ponderosa* Dougl.

Plantaginaceae, Plantain Family
Common Plantain   *Plantago major* L.
English Plantain   *P. lanceolata* L.

Polygonaceae, Buckwheat Family
Bitter Dock   *Rumex obtusifolius* L.
Western Dock   *R. occidentalis* Wats.
Sheep Sorrel   *R. acetosella* L.
Knotweed   *Polygonum aviculare* L.
Japanese Knotweed   *P. cuspidatum* Sieb.&Zucc.

Polypodiaceae, Fern Family
Brake F.   *Pteridium aquilinum* (L.)Kuhn
Deer Fern   *Blechnum spicant* (L.)Roth.
Lady Fern   *Athyrium felix-femina* (L.)Roth.
Licorice Fern   *Polypodium glycyrrhiza* DC.
Maidenhair Fern   *Adiantum pedatum* L.
Sword Fern   *Polystichum munitum* (Kaulf.) Presl.
Wood Fern   *Dryopteris austriaca* (Jacq.)Wagner

Portulacaceae, Purslane Family
Candy Flower   *Montia sibirica* (L.)Howell
Miner's Lettuce   *M. perfoliata* (Donn.)Howell

Primulaceae, Primrose Family
Starflower   *Trientalis latifolia* Hook.

Ranunculaceae, Buttercup Family
Baneberry   *Actaea rubra* (Ait.) Wild.
Bugbane   *Cimicifuga elata* Nutt.
Cr.Buttercup   *Ranunculus repens* L.
Little Woods B.   *R. uncinatus* D.Don.

Western B.   *R. occidentalis* Nutt.
W.Clematis   *Clematis ligustifolia* Nutt.
W. Red Columbine   *Aquilegia formosa*
W. Meadowrue   *Thalictrum occidentale* Gray
Windflower   *Anemone deltoidea* Hook.

Rhamnaceae, Buckthorn Family
Buckbrush   *Ceanothus sanguineus* Pursh
Cascara   *Rhamnus purshiana* DC.

Rosaceae, Rose Family
Yel.Avens   *Geum macrophyllum* Willd.
EvergreenBl.   *Rubus laciniatus* Willd. Weihe & Nees
Bit.Cherry   *Prunus emarginata* (Dougl.)Walp.
Blackcap   *Rubus leucodermis* Dougl.
Crabapple   *Pyrus fusca* Raf.
Goatsbeard   *Aruncus sylvester* Kostel
Hardhack   *Spiraea douglasii* Hook.
Black Hawthorn   *Crategeus douglasii* Lind.
Red Hawthorn   *C. aucuparia* L.
Indian Plum   *Oemleria cerasiformis* (H.&A.)Landon
Laurel Cherry   *Prunus laurocerasis* L.
Mountain Ash   *Sorbus aucuparia* L.
West. Mt. Ash   *S. sitchensis* Roemer
Ocean Spray   *Holodiscus discolor* (Pursh)Maxim.
Nootka Rose   *Rosa nutkana* Presl.
Clustered Wild Rose   *R. pisocarpa* Gray
Sweetbriar Rose   *R. eglanteria* L.
Wood Rose   *R. gymnocarpa* Nutt.
Beach Str.   *Fragaria chiloensis* (L.)Duchesne
Wood Strawberry   *F. vesca* L.
Salmonberry   *Rubus spectabilis* Pursh
Service Berry   *Amelancier alnifolia* Nutt.
Thimbleberry   *Rubus parviflorus* Nutt.

Rubiaceae, Madder Family
Bedstraw   *Galium aparine* L.
Fragrant Bedstraw   *G. triflorum* Mich.
Sweet Woodruff   *Asperula odorata* L.

Salicaceae, Willow Family
Black Cottonwood   *Populus trichocarpa* T.&G.
Piper's Willow   *Salix piperi* Bebb
Long-Leaf Willow   *S. lasiandra* Benth.
Scouler Willow   *S.scouleriana* Barratt

Saxifragaceae, Saxifrage Family
Alumroot   *Heuchera micrantha* Dougl.
Bishop's Cap   *Mitella caulescens* Nutt.
Foamflower   *Tiarella trifoliata* L.

Fringecup   *Tellima grandiflora* (Pursh)Dougl.
Piggy-Back Pl.   *Tolmiea menziesii* (Pursh)Dougl.
Scrophulariaceae, Figwort Family
Calif. Figwort   *Scrophularia californica* Cham. & Schecht
Foxglove   *Digitalis purpurea* L.
Yellow Monkey Fl.   *Mimulus guttatus* DC.
Monkey Flower   *M. dentatus* Nutt.
Monkey Flower   *M. moschatus* Dougl.
Common Mullein   *Verbascum thapsus* L.
Amer. Speedwell   *Veronica americana* Schwein
Solanaceae, Potato/Nightshade
Bittersweet Nightshade   *Solanum dulcamara* L.
Deadly Nightshade   *S. nigrum* L.
Taxaceae, Yew Family

Pacific Yew   *Taxus brevifolia* Nutt.
Typhaceae, Cat-tail Family
Broad-leaved Cat-tail   *Typha latifolia* L.
Umbelliferae (Apiaceae), Parsley
Cow Parsnip   *Heracleum lanatum* Michx.
Poison Hemlock   *Conium maculatum* L.
Queen Anne's Lace   *Daucus carota* L.
Water Parsley   *Oenanthe sarmentosa* Presl.
W. Sweet Cicely   *Osmorhiza chilensis* H.&A.
Urticaceae, Nettle Family
Stinging Nettle   *Urtica dioica* L. var.*lyallii*
Violaceae, Violet Family
Evergreen V.   *Viola sempervirens* Greene
Purple Violet   *V. adunca* Sm.
Wood Violet   *V. glabella* Nutt.

Western Wahoo
Euonymus occidentalis

PEOPLE

Aristotle, 384-322 B.C. A Greek philosopher, biologist, psychologist, political thinker, whose books contributed much to the advancement of European learning.

Avicenna, (Ibn-Sina), 980-1037, Arab philosopher and physician, one of many who preserved Greek and Roman plant knowledge during the Dark Ages. His *Canon of Medicine* included doctrines of Galen, Hippocrates, and Aristotle. An 1187 translation of Avicenna was the standard textbook of medicine for more than 400 years.

Bailey, Liberty Hyde, 1858-1954, American teacher, horticulturalist, and botanist who wrote and edited many books. Author of Hortus Third, considered the authority on botanical names of horticultural plants. Abbreviation used in scientific names: Bail.

Bentham, George, 1800-1884, British botanist associated with Sir Joseph Dalton Hooker at Royal Botanic Gardens at Kew. His *Genera Plantarum*, 1883, described 202 families placed in orders. Abbreviation, Geo. Bentham, Benth.; Bentham and Hooker, B.&H.

Bessey, Chas. Edwin, 1845-1915, taught botany in United States from 1871 and wrote many books. He is best known for the phylogenetic system he presented in 1898 and 1915. His practices are still considered significant, his dicta highly regarded; his concepts form the foundation of the most widely-accepted modern systems.

Bock, Hieronymus, (Jerome), 1498-1554. A Lutheran pastor and botanist whose *Kreuter Buch,* 1539, written from his own experience, included German plants. He showed similarities of relationships, classified plants as herbs, shrubs, and trees.

Brown, Robert, 1772-1858, Scottish botanist and naturalist associated with the British Museum. He proved that gymnosperms, having naked ovules and seeds, were a separate group. He worked toward establishing larger taxa now known as orders.

Camerarius, Joachim, the younger, 1534-1598, German botanist, credited with establishing the fact of sexuality of flowering plants. His herbal *Hortus Medicus* was published in 1588.

de Candolle, Augustin Pyrame, 1778-1841, Swiss botanist, outstanding for classification. His *Theorie elementaire de la botanique,* 1813, included 135 families. An 1844 publication by his son **Alphonse Louis Pierre Pyrame de Candolle,** 1806-1893, had 213 families. Grandson Casimir was also involved in botany. Abbreviations: DC.; son Alphonse, DC.A.

Clark, William, 1770-1838, American soldier and explorer; with Meriwether Lewis he commanded the Lewis and Clark expedition of 1803-06 which went from St. Louis, Missouri, to the mouth of the Columbia river and return.

Clusius, Carolus, (Charles de l'Ecluse), 1526-1609, a leading European botanist in 1580s; worked in Imperial Gardens in Vienna; associate of Dodoens and de Lobel. Published *Rariorium Plantarum Historia* in 1601.

Colonna, Fabio, 1567-1651, born in Naples, contributed to concepts of genera and species. He published two major botanical works illustrated by his own etchings.

Cratevas, (Krateuas), 122-63 B.C., Greek herbalist, physician, and artist, considered the father of botanical illustration. Some of his designs were used in the earliest-known surviving herbal of Dioscorides, a copy from 512 A.D.

Cronquist, Arthur, contemporary botanist, senior scientist with N. Y. Botanical Garden, one of the authors of the 5-volume *Vascular Plants of the Pacific Northwest.* He published in 1968 *The Evolution & Classification of Flowering Plants,* a major work about the phylogenetic order of plants.

Culpeper, Nicholas, 1616-1654, English herbalist. In 1649 he translated from Latin to English, the London Physical Directory, source of medicinal information for doctors; published complete herbal, *The English Physitian,* 1653.

Darwin, Chas. Robert, 1809-1882, English naturalist. In his *Origin of Species,* 1859, he noted variation in living things, emphasized effect of natural selection upon survival. Darwin left a bequest for a list of known flowering plants, which became the Index Kewensis.

Dioscorides, Pedianos, c. 40-100 A. D., Greek doctor in the Roman army under Nero. His *De Materia Medica,* frequently cited in European herbals, included medicinal properties of about 600 plants, mostly Mediterranean. Forty-four drugs included by Dioscorides were still in European pharmacopoeias in the 1920s.

Dodoens, Rembert, (Dodonaeus), 1517-1585. Dutch physician and botanist. His 1554 *Cruydeboeck,* illustrated with 1300 woodcuts, dealt with plants of the Netherlands. It was translated to English by Henry Lyte as *Nieuve Herball or Historie of Plantes,* and is considered the source of plant information used by Edmund Spenser and Shakespeare. His work was also copied extensively in Gerard's *Herball* and other herbals.

Douglas, David, 1798-1834, Scottish naturalist and botanist. During trips to the Pacific Coast from 1825 to 1833, he discovered more than 50 species of trees, more than 100 species of shrubs, ferns, and other plants. He introduced many western plants into Britain. Local plants honoring him include Hardhack (*Spiraea douglasii),* and Douglas Fir. Abbrev: Dougl.

Engler, Heinrich Gustav Adolph, 1844-1938, German botanist. His phylogenetic system, published from 1877 to 1899 in *Die Naturlicher Pflanzenfamilien,* was widely accepted. Hitchcock et al classified dicotyledons in the traditional Englerian sequence.

Fuchs, Leonhart, 1501-1566, German physician, teacher, and botanist. His herbal, *De Historia Stirpium,* 1542, had excellent illustrations, extensively copied in later herbals, some used as types by Linnaeus. He studied the number of stamens and styles in flowering plants.

Galen, c.130-c.200, Greek physician and writer who settled in Rome. Next to Hippocrates, he is considered the most distinguished physician of antiquity. His writings included plant information.

Gerard, John, 1545-1612, English botanist, herbalist, surgeon, who listed 1039 plants in his garden in 1596. His *Herball or Generall Historie of Plantes,* 1597, was basically a reproduction of Dodoens in English. The 1800 illustrations in the first edition were from stock woodcuts, many identified by de Lobel. In 1633, Thomas C. Johnson, 1597-1644, published a much-improved edition of Gerard's *Herball* which is still popular.

Goodyer, John, 17th century English botanist, translated Dioscorides into English in 1655.

Gray, Asa, 1810-1888, American botanist, taxonomist and teacher. Dr. Gray worked with discovery and classification of North American plants. He wrote many botanical texts and scientific papers, including, with John Torrey, *A Flora of North America.* ID: Gray; with Torrey: T.&G.

Herodotus c.484 B.C.-c.425 B.C. Greek historian and explorer. His writings provide an early record of plants which had been cultivated for a long time.

Hippocrates, 460 B.C.-377 B. C. Greek Physician, called the father of medicine. He is said to have separated medicine from philosophy. His writings gave medicinal properties of 300 to 400 plants.

Hooker, Sir Joseph Dalton, 1817-1911, British botanist, director of the Royal Botanic Gardens, Kew, 1865-85; an associate of George Bentham. He visited the American Rockies in 1877. ID: Hook.f. (f. is the abbreviation of the French *fils,* son). With Bentham, ID: B.& H.

Hooker, Sir William Jackson, 1785-1865, English botanist, first director of the Royal Botanic Gardens at Kew, father of Sir J. D. Hooker. Published *Flora Boreali Americana,* 1829 to 1834. ID: Hook.

Howell, Thomas Jefferson, 1842-1912. A self-taught Oregon pioneer who collected and classified the most complete list of plant life of Oregon and Washington prepared up to his time. He published this himself, the last part in 1903, as *Flora of Northwest America.*

de Jussieu, Bernard, 1699-1777, one of a French family distinguished for botanists. He left a list which became the basis for family names in the International Code of Botanical Nomenclature. ID: Juss.B. His nephew,

de Jussieu, Antoine Laurent, 1748-1836, French botanist and teacher, wrote *Genera Plantarum,* 1789, considered the foundation of modern classification. ID: Juss.A.

Lewis, Meriwether, 1774-1809, American explorer and co-leader with William Clark of the first overland expedition to the Pacific, which brought back valuable scientific information.

Linnaeus, Carolus, 1707-1778, Swedish naturalist whose name, **Carl Linne,** was Latinized in the custom of his time. A systematist, he is famous for publishing a logical classification system which established principles for defining genera and species. His *Species Plantarum,* 1753, is the starting point of genus names of flowering plants; about 5900 species were then known. Though he was not the first to suggest the binomial system, his consistent use of it was a great contribution. ID: L.

de Lobel, Matthias, (Lobelius, de L'Obel), 1538-1616, Belgian/French physician. His 1581 publication, *Kruydtboek,* attempted a system of classification by relationships.

Lyall, David, 1817-1896, Scottish botanist, naturalist and doctor of medicine, collected in North America while serving the North American boundary commission during the mid-19th century.

Magnol, Pierre, 1638-1715, professor of medicine and botany at Montpellier, France. He developed the concept of families as having striking characteristics of roots, stems, flowers, and seeds. His 1689 publication listed 76 families.

Mendel, Gregor Johann, 1822-1884, Austrian monk and botanist, in 1865 published studies concluding that inherited variations can be passed on, those due to environment cannot be passed on.

Menzies, Archibald, 1754-1842, Scottish physician and botanist who came to the Northwest with Captain George Vancouver expedition of 1790-95. He reported finding numerous plants and trees; names showing his influence are Piggy-back plant, *Tolmiea menziesii,* Douglas Fir, *Pseudotsuga menziesii,* Madrona, *Arbutus menziesii.* ID: Menz.

Michaux, Andre, 1746-1802, French botanist who came to North America in 1785. Among his books were one on North American Oaks in 1801, and the two-volume *Flora Boreali Americana,* 1803. His son Francois published from 1810 to 1813 a three-volume book on North American forest trees. ID: Michx.

Nuttall, Thos. 1786-1859, English-born botanist and ornithologist, who published, 1818, *The Genera of North American Plants.* He collected botanical specimens in the Pacific Northwest in 1834-35. ID:Nutt.

Paracelsus, Philippus Aureolus, (real name, Theophrastus Bombastus von Hohenheim), c.1490-1541. Swiss-born alchemist and physician; his name is associated with the Doctrine of Signatures. (See Heal-all.)

Parkinson, John, 1567-1650, English apothecary, herbarist to Charles I and James I. His *Theatricum Botanicum,* 1640, (Theater of Plants,) *an Herball of Large Extent,* had 1,755 folio pages, 2700 woodcuts, described 3800 plants. It was one of the last to use woodcut illustrations. He introduced 7 plants to England, provided first mention of 33 English native plants.

Pliny, Caius Plinius Secundus, (Pliny the Elder), 23-79 A. D., Roman naturalist. About half of his *Naturalis Historia* dealt with plants and medical botany. Pliny was much cited in medieval herbals.

Pursh, Frederick Traugott, 1774-1820, born in Siberia, educated in Germany, made important contributions to North American botany. He published *Flora Americae Septrionalis* (northern), an early flora, in 1814. ID: Pursh.

Ray, John, 1627-1705, British naturalist and minister. *Historia generalis plantarum,* three volumes published in 1686, 1688, and 1704, established division of flowering plants into dicots and monocots.

Scouler, Dr. John, 1804-1871, Scottish surgeon and naturalist for the Hudson's Bay Company, who explored the Northwest coast of North America with David Douglas in 1825. Bellflower, *Campanula scouleri,* Scouler willow, *Salix scouleriana.*

Takhtajan, Armen, contemporary Russian plant systematist. His *Flowering Plants—Origin and Dispersal,* 1969, is a major work on the phylogenetic order of plants.

Theophrastus, c.372 B.C.-287 B.C., Greek naturalist and philosopher. His work was the beginning of our surviving written plant history, and the most important contribution to botanical science during antiquity and the Middle Ages. His *Enquiry into Plan's* included 300 species, mostly cultivated plants used medicinally. Concerned largely with eastern Mediterranean flora, he was aware—unlike most people of his time—of differences in plants growing in different regions.

Tolmie, Dr. William Fraser, 1812-1886, Scottish doctor, went in 1832 to Hudson's Bay Company's Fort Vancouver. His interest in botany is reflected in plants bearing his name, one being the familiar Piggy-Back Plant, *Tolmiea menziesii.*

Torrey, John, 1797-1873. American systematic botanist, teacher, author, who wrote, in collaboration with Asa Gray, *A Flora of North America,* 1840. ID: Torr.; with Gray: T.& G.

de Tournefort, Josephus Pitton, 1656-1708; French botanist, pupil of Magnol, published *Institutiones Rei Herbariae* in 1700. He is given credit for the concept of genera.

Turner, Wm. 1500-1568, English botanist, physician and chaplain. He differentiated classical, continental, and English species, an advanced idea for his time. English botany is said to begin with his three-part *New Herball,* published in 1551, 1562, and 1568, which included over 200 native species and had excellent new illustrations.

Fragrant Bedstraw
Galium triflorum
piece showing whorled leaf arrangement

BOOKS

Books are a changing and enjoyable resource; each has benefits and drawbacks. Flower books often omit woody plants—shrubs—which excludes such "flowers" as wild Roses. Some limit contents to native plants, eliminating Dandelion, Foxglove, Plantains, and other well-established foreigners. Field guides may have lovely illustrations, nice symbols for plant families—and cover a tremendous area. Wide coverage leaves out plants not representative of the larger area, and includes some enough like local species to be confusing. When using books first published earlier, it is good to be aware that scientific names do change. Checking synonyms may tell whether information in an older book is about a plant you know or seek. I like to identify plants by pictures—which isn't always easy! Pictures close to descriptions are often an advantage of black and white illustrations.

The area covered is important. Native plants common west of the Rocky Mountains do not normally grow wild in the east, and the reverse is also true; different species developed on both sides of the barrier created thousands of years ago by the uplifting of the Rockies. Many books about Pacific Northwest wildflowers have been published in the past ten years. Books are a joy—a frustration—a tool! Following are some I especially like.

Handy-Dandies, Basically Northwest

Haskin, Leslie L., *Wild Flowers of the Pacific Coast*, originally published in 1934. Includes many stories about plants. Dover Publications, Inc., New York

Hill, Clara Chapman, *Spring Flowers of the Columbia River Valley*, U. of W. Press, 1958. Includes uses and has excellent black and white drawings close to plants.

Horn, Elizabeth L., *Wildflowers I, the Cascades*, and *Wildflowers, the Pacific Coast*, Beaverton, OR, 97005, are two of several books Mrs. Horn has written about plants in our area; beautiful photographs, identifying and miscellaneous information.

Larrison, Patrick, Baker and Yaich, *Washington Wildflowers*, Seattle Audubon Society, 1974. Common Wildflowers; no introduced plants, trees, or shrubs. Colored and black and white illustrations, maps showing environmental diversity, photographic suggestions, local orientation.

Lyons, C. P., *Trees, Shrubs and Flowers to know in Washington*, J.M. Dent & Sons, Toronto; 1975. Including trees, shrubs and ferns is an advantage. Black and white illustrations, plant identification and uses. This is my old standby.

Robinson, Peggy, *Profiles of Northwest Plants*, 1979; Far West Book Service, 3515 N. E. Hassalo, Portland, OR, 97232. Plant uses, including Indian, medicinal, and literary references. Informative and entertaining.

Edible Plants

Books about edible plants range from the many popular books by Euell Gibbons, Bradford Angier, and others to more specifically local presentations. I have enjoyed:

Harrington, M. C., *Edible Native Plants of the Rocky Mountains*, U. of New Mexico Press, 1967. Many recipes, lengthy report on Elderberries.

Hedrick, U. P., *Sturtevant's Edible Plants of the World*. Dover Publications, Inc., New York. Encyclopedic notes made from 1867 to 1896 by Edward Lewis Sturtevant, director of the N. Y. Ag. Exp. Station; first published 1919. No illustrations.

Kirk, Donald, *Wild Edible Plants of the Western United States*, Naturegraph Publishers, Healdsburg, CA, 95448, 1970. A field guide with black and white illustrations close to plant descriptions; food uses, other information.

Mohney, Russ, *Why Wild Edibles,* Pacific Search, 1975. Covers our area, includes recipes, identification; illustrations close to plant descriptions.

Indian Plant Uses

Balls, Edward K., *Early Uses of California Plants,* 1962; No. 10 of California Natural History Guides, University of Cal. Press.

Dickson, Evelyn, *Food Plants of Western Oregon Indians,* unpublished MA thesis, School of Education, Stanford University, 1946. On microfilm in the library of the Oregon Historical Society, Portland.

Gunther, Erna, *Ethnobotany of Western Washington,* U. of W. Press, 1973; the best available source for plant uses by Western Washington Indians.

Murphey, Edith Van Allen, *Indian Uses of Native Plants,* 1959, Mendocino County Historical Society, Fort Bragg, CA, 95437. Mrs. Murphey worked for the Bureau of Indian Affairs for many years.

Sweet, Muriel, *Common edible and useful Plants of the West,* Naturegraph Company, 1962. Many California plants.

Turner, Nancy J., *Food Plants of British Columbia Indians, Part 1, Coastal Peoples, Part 2, Interior Peoples,* and *Plants in British Columbia Indian Technology,* are numbers 34, 36, and 38 in a reasonably-priced nature series. These detailed books give plant uses, methods of preparation, seasonal information and much more. If not available locally, write PUBLICATIONS, British Columbia Provincial Museum, Victoria, B.C., V8V 1X4.

Vogel, Virgil J., *American Indian Medicine,* U. of Oklahoma Press, 1970. A technically oriented book giving recorded medicinal use by Indians of central and eastern United States, of native plants, identified by genus and species.

Medicinal uses of Plants

Grieve, Mrs. M., *A Modern Herbal, the Medicinal, Culinary, Cosmetic and Economic Properties, Cultivation and Folklore of Herbs, Grasses, Fungi, Shrubs, and Trees with all their Modern Scientific Uses.* Dover Publications, Inc., New York, 1971. Plants used for herbal medicine with information from literature, history, folklore, botany.

Lewis, Walter H. & Memory Elvin-Lewis, *Medical Botany, Plants affecting Man's Health,* John Wiley & Sons, 1977. A college text; extensive lists of medical uses for plants.

Moore, Michael, *Medicinal Plants of the Mountain West,* 1979, Museum of New Mexico Press, P.O. Box 2087, Santa Fe, New Mexico, 87503. A current herbal.

Osol, Arthur, and George E. Farrar, Jr., *The Dispensatory of the United States of America,* 1950 edition, Philadelphia, J. B. Lippincott Co., 1947-50. A weighty tome designed for pharmacists which includes information about plants which have had medicinal uses.

Stuhr, Ernst T., *Manual of Pacific Coast Drug Plants,* 1933. Dr. Stuhr was an Associate Professor of Pharmacology and Pharmacognosy at Oregon State. Local plants.

Poisonous Plants

Dennis, LaRea J., *Name your Poison, A Guide to Cultivated & Native Oregon Plants Toxic to Humans,* 1972, Department of Botany and Plant Pathology, O.S.U. An amazingly-complete series of plant lists.

Kingsbury, John M., *Deadly Harvest,* Holt, Rinehart and Winston, New York, 1965. Plant poisons and the reactions they cause.

Muenscher, W. L., *Poisonous Plants of the United States,* 1957, The Macmillan Company, N. Y. Descriptions, identification, black and white illustrations, conditions of poisoning.

Other

Adrosko, Rita J., *Natural Dyes and Home Dyeing*, Dover Publications, Inc., New York

Clark, Lewis J., *Wild Flowers of the Pacific Northwest*, Gray's Publishing, Ltd., Sidney, B. C., 1976, has outstanding photography, comfortably-presented botanical information and a wealth of miscellany. Six field guides were prepared from this large volume. These are again available, (1985), from the U. of W. Press, Seattle, WA, 98145. I lists flowers of Forest and Woodland; II, Field and Slope; III, Marsh and Waterway; IV, Coast Flowers; V, Arid Flatlands; VI, Mountain Wildflowers.

Fry, Theodore C. *Ferns of the Northwest*, originally published 1934, now Dover, New York.

Hitchcock, C. Leo, A. Cronquist, M. Ownbey, and W. Thompson, *Vascular Plants of the Pacific Northwest*, U. of W. Press, 1955-1969, 5 vols. This, the authority used here, gives extensive botanical information and black and white drawings for each plant. It lists synonyms. Some basic knowledge is essential before approaching either the five-volume set or

Flora of the Pacific Northwest, C. Leo Hitchcock & Arthur Cronquist, U. of W. Press, 1979. This one-volume condensation of the large set is a college text. It has illustrations, botanical identification, and is (almost) small enough to carry in the field. Space-saving abbreviations make this harder to use than the unabridged edition.

Kozloff, Eugene N., *Plants and Animals of the Pacific Northwest, An Illustrated Guide to the Natural History of Western Oregon, Washington, and British Columbia*, U. of W. Press, 1976. This locally-oriented volume has both color and black and white illustrations, emphasizes habitats and points out interrelationship of plants and animals. Easy to read and understand.

Ross, Charles R., *Ferns to know in Oregon*. Ore. Ext. Service Bulletin 785, O.S.U.

Ross, Charles R., *Trees to know in Oregon*, Ore. Ext. Service Bulletin 697, O.S.U.

Dogwood
Cornus nuttallii

GLOSSARY

achene, akene A small, dry, one-seeded fruit which does not open when mature; Maple, Clematis. Gr. *a*, not, *chainein*, to gape.

alkaloid A complex basic or alkaline compound containing nitrogen, usually from plants, usually bitter, usually potentially toxic. Alkaloids occur almost exclusively in seed-bearing plants, are rarely found in plants reproduced by spores. Individual plants contain more than one alkaloid; function in plants is uncertain. These are widely used in drug and herbal therapy. Examples: caffeine, morphine, strychnine. Arabic, *alqili*, ashes of the plant Saltwort; Gr. *eidos*, form.

allergen A substance causing an exaggerated reaction or abnormal sensitivity. Gr. *allos*, different, atypical, *genein*, to produce.

alterative Substance which gradually improves health. L. *alterare*, to change.

analgesic Substance which relieves pain. Gr. *a*, not, *algeein*, feel pain.

angiosperm Flower-bearing plant having ovules enclosed in an ovary that forms the fruit after fertilization. Gr.*angeion*, vessel, *sperma*, seed.

annual A plant completing its life span in one growing season. L. *annualis*, within a year. A biennial lives for two years.

anthelmintic See vermifuge. *anti*, against, Gr. *helminth, helmis*, worm.

anther The part of the stamen which develops and contains pollen. Gr. *anthemon*, flower.

antiscorbutic A remedy for scurvy, a disease caused by lack of fresh vegetable food. Gr. *anti*, against, L. *scorbutus*, scurvy.

antiseptic Preventing infection. Gr. *anti, sepein*, to rot.

antispasmodic An agent relieving or preventing spasms. Gr. *anti, spaein*, to draw up, tear away.

aperient A gentle laxative. L. *aperire*, to open.

astringent Substance that shrinks tissues, used to stop bleeding, etc. L. *astringere*, to draw tight.

axil The angle between the upper side of a leaf and the stem from which it grows. Gr. *axilla*, armpit.

berry A soft fruit with seeds in pulpy tissue, as grape, date, tomato, orange. Anglo-Saxon *berie, berige*.

blade The flat wide part of a leaf or flower petal. Anglo-Saxon *blaed*.

bract A small leaf near the base of a flower, flower cluster or spore case. L. *bractea*, a thin metal plate.

bulb A short underground storage organ with overlapping, enlarged leafy scales, such as lily, onion, tulip. Gr. *bolbos*, onion, L. *bulbus*, globular root.

calyx The outer series of floral leaves, usually green, individually called sepals. Gr. *kalyx*, a husk, cup.

carminative Medicine to expel gas from stomach, intestines, or bowels. L. *carminare*, to card wool, hence, to cleanse.

cathartic A substance which causes emptying of the bowels. Laxatives or aperients are mild cathartics; purgatives, powerful cathartics. Gr. *katharos*, to cleanse.

catkin A flowering spike which eventually falls from the plant entire; typical of Willows. The small flowers are unisexual, have many bracts, and are usually without stems. An ament. Dutch *katteken*, little cat.

chlorophyll Green plant pigment which absorbs light energy in the process of photosynthesis. In the presence of light, it makes carbohydrates such as starch from carbon dioxide and water. Gr. *chloros*, green, plus *phyllon*, leaf.

compound leaf A leaf divided into two or more parts or leaflets. L. *com,* with or together, *ponere,* to put.

cone A fruiting organ with modified leaves, (scales), which bear pollen sacs or ovules, often arranged in a spiral, as a Pine cone. A strobile. Some sources limit use of "cone" to gymnosperms; others include somewhat similar structures such as "cones" of Alder. Gr. *konos,* a Pine cone.

conifer A cone-bearing tree. L. *conus,* cone; *ferre,* to bear.

corm A short, solid, vertical, enlarged underground stem in which food is stored. Gr. *kormos,* a trunk.

corolla The flower petals, sometimes united into a tube. L. *corolla,* diminutive of *corona,* a wreath.

counterirritant An irritant substance used externally to help heal by increasing blood circulation: mustard plasters. L. *contra,* against, *irritare.*

cyme A cluster in which flowers at the top of the stalk open before those below. Gr. *kyma,* a wave, a swelling.

deciduous Falling in season, as petals after flowering or leaves in the autumn. L. *de,* from, *cadere,* to fall.

decoction A solution with water made by boiling slowly for a long period; hard substances such as bark, large seeds and nuts, usually require heat for a long time to extract their virtues. L. *de, coquere,* to cook, boil.

dehiscent/indehiscent Splitting when ripe; indehiscent, not opening when ripe. L. *dehiscere,* to split open, to divide.

demulcent A medicine used internally to soothe or protect mucous membrane from irritation. L. *de, mulcere,* to sooth.

dermatitis Skin inflammation or disease. Gr. *derma,* skin, *-itis,* disease.

diaphoretic A substance which increases or promotes perspiration. Gr. *dia,* through, *pherein,* to carry.

dioecious Having male and female parts on different plants; flowering plants with staminate and pistillate flowers on separate plants and conifers having pollen and seed cones on different individuals. Gr. *dis,* twice, *oikos,* house.

dispensatory A book of systematic descriptions of drugs. L. *dispensare,* to weigh out, dispense.

diuretic A substance increasing flow of urine. Gr. *dia,* through, *oureein,* to urinate.

drupe Fruit with a fleshy outer part and a hard center or pit which contains a seed; cherry or raspberry, etc. Gr. *drypeps,* ripening on the tree.

emetic A substance inducing vomiting. Gr. *emetikos,* to vomit.

emmenagogue A substance promoting menstrual flow. Gr. *menses,* month.

emollient A soothing ointment used externally. L. *e,* out of, *mollire,* to soften.

expectorant Medicine which helps remove mucus from lungs and throat. *ex,* out of, from, *expectus, expectoris,* the breast.

febrifuge Agent reducing fever. L. *febris,* fever, *fugere,* to put to flight.

formulary A book of prescribed formulae, especially medicinal compounds. L. diminutive of *forma,* form.

fruit Plant part containing seed and other parts developing with it. L. *fructus,* that which is enjoyed.

genus Classification unit (taxon) between family and species; L. *genus,* birth, race, kind.

glabrous A plant surface without hairs, smooth. L. *glaber,* bald.

gland A small organ that secretes oil or nectar. L. *glano, glandis,* acorn.

glycoside / glucoside Gr. *glykys*, sweet, the name source, reports that one component converts into the sugar glucose. (Gr. *eidos*, form.) Glycosides contain several sugars. Glucose is not a poison but other plant compounds may interact with body substances to form combinations poisonous to the system.

Several types of glycosides include but are not limited to the following: 1. Cardiac glycosides are most important therapeutically; one of these is Foxglove, *Digitalis purpurea*. 2. Glycosidal anthraquinones are purgatives; purgative vegetable drugs are dependent at least in part on glycosides. Cascara, rhubarb and senna fit this classification. 3. Saponins (which see), yield various sugars. English Ivy, *Hedera helix*. 4. Cyanogenic glycosides, widely distributed in nature, yield hydrocyanic acid. Poisoning results from release of the acid by digestive enzymes of men or animals. Bitter Almonds, Cherry Laurel, and Elderberry all have this. 5. Glycosidal dyes and pigments, one of which is quercitrin, present in Oak bark.

head A flower arrangement typical of the Composite family, consisting of a crowded cluster of stemless flowers at the top of a single stem or peduncle, as in Clover and Daisy. Anglo-Saxon *heafod*.

hemostatic A substance which can stop bleeding. Gr, *haemo*, blood, *histasthai*, to stand.

herb A seed plant which does not develop woody tissue above ground, having above-ground parts which die at the end of a season's growth. "Herbs" used for seasoning food or preparing medicine are sometimes parts of shrubs or trees. L. *herba*, grass, green blades.

inflorescence The flower cluster or the manner of arrangement of flowers. L. *inflorescere*, to begin to bloom.

infusion An extract made by steeping or soaking in water or other liquid without boiiing; a tea. L. *infusare*, to pour in.

involucre A whorl of bracts around a flower cluster. L. *involucrum*, a wrapper.

leaflet A single part of a compound leaf, diminutive of leaf.

lenticel See stoma. L. *lens, lentis*, a lentil.

lobe A rounded leaf division, cut less than halfway to the midrib. Gr. *lobos*, lower part of the ear.

monoecious Seed plants having separate reproductive parts both borne on the same plant. Gr. *monos*, single, *oikos*, house.

node The slightly larger part of a stem where leaves or branches originate. L. *nodus*, a knot.

nut A dry, hard, one-seeded fruit which does not open when ripe. L. *nux*, a nut; Anglo-Saxon, *hnutu*.

official A medication sanctioned by the pharmacopoeia. L. *officialis*, from *officium*, service.

officinale A medication kept in stock by druggists, recognized by the pharmacopoeia. L. *officina*, workshop.

order Next larger taxon than family.

organ A part adapted for a particular function, as leaf, root, or flowers of a plant. Gr. *organon*, from *ergon*, work.

ovary / ovule The enlarged lower part of the pistil, within which seeds develop; the immature seed L. *ovum*, an egg, *ovulum*, diminutive of *ovum*.

palmate Radiating from a central point like the fingers of a hand. L. *palma*, palm of hand.

parasite An organism getting its food from the living parts of another plant or animal. Gr. *parasitos*, one who eats at the table of another.

pectoral Relating to the lungs or the chest organs. L. *pectus, pectoris*, the breast.

pedicel Stem of a single flower in a flower cluster. L. *pediculus*, a little foot.

peduncle The stem of a single flower, or the main stem of a flower cluster. L. *pedunculus*, diminutive of *pes*, foot.

perennial A nonwoody plant having underground parts that live for more than two years. L. *per* through, *annus*, year.

petal A unit of the corolla. Gr. *petalon*, a flower leaf.

petiole Stem of a leaf. L. *petiolus*, a little foot or legs.

pharmacopoeia A book describing drugs and medicinal preparations. Gr. *pharmakon*, drug, *poieein*, *make*.

photosynthesis The process by which carbon dioxide and water are brought together chemically to form a carbohydrate, using radiant energy. Gr. *photos*, light, *syn*, together, *tithenai*, to place. See chlorophyll.

phylogenetic The evolutionary history of a group of organisms, plants for example. Gr. *phylon*, race or tribe, *geneia*, origin.

phylum/phyla A primary division of the animal or vegetable kingdom; from *phylon*, race or tribe, *geneia*, origin.

pinnate Featherlike, having parts arranged on two sides of a midrib. L. *pinnatus*, feathered.

pistil The seedbearing organ of a flower. L. *pistillum*, a pestle.

pod A dry seedcase, splitting top to bottom along two sides at maturity, as in the pea and bean; a dry fruit which splits when ripe. Gr. *pes*, foot.

pollen microspores of higher plants, (seed plants), usually a fine yellow dust. The male gametophyte that develops sperms. L. *pollon*, fine flour or dust.

pome A core fruit characteristic of the apple. L. *pomum*, apple.

potherb Plant having leaves and stems which are boiled for food, as spinach.

purgative See cathartic. L. *purgare*, to purify, purge.

raceme A long unbranched flowering stalk with flowers commencing to bloom at the bottom, individual flowers having stems about the same length. L. *racemus*, a bunch of grapes.

rhizome An elongated, rootlike, underground horizontal stem, producing leaves on the upper side and roots on the lower; a rootstock. Gr. *rhiza*, root.

rubefacient Medicine used externally which causes redness and increased blood supply; a counterirritant. L. *rubefacere*, to make red.

runner A stem which grows along the ground, often developing new plants at nodes or tips; a stolon. Anglo-Saxon *rinnan*, run.

samara Dry winglike fruit with one or two seeds which do not open, as those of the Ash, Maple, Elm; a key or key fruit. L. *samara*, the fruit of the Elm.

saponin A glycoside occurring naturally in many plants, which lather when crushed and rubbed in water. Saponins have various sugars; they are reported to occur in 500 genera of plants. Eating too much can cause diarrhea; saponins can often be boiled out of a plant, making it more edible. L. *sapo*, *saponis*, soap.

saprophyte An organism taking its food from the dead body or the nonliving products of another plant or animal. A saprophytic plant lacks chlorophyll and cannot manufacture its own food. Gr. *sapros*, rotten, *phyton*, a plant.

scales Petal-like parts of a cone, or modified underground leaves of bulbs. Old French *escale*, cup.

seed A mature ovule; Anglo-Saxon *saed*.

sepal Short for Latin *separatum petalum*, separate petal. See calyx.

sessile Without a stalk, as a leaf attached directly to a stem. L. *sessilus*, sitting, from *sedere*, to sit.

206

shrub A perennial woody plant of relatively low stature with several stems arising from or near the ground. Old English *scrybb*, brush.

species A group of individuals with many characteristics in common, which usually interbreed freely. Next smaller classification unit than genus; abbreviation: spp. Species is both singular and plural. L. *species*, sort, form, kind.

spike A long flower cluster having stemless flowers; a raceme has flowers with stems. L. *spica*, an ear of grain.

spore A primitive reproductive body able to develop into a new individual. The lower and lower vascular plants, some of which are Mosses, Lichens, Fungi, Ferns and Horsetail, reproduce by spores. Gr. *spora*, seed.

stamen Male organ of flower structure. L. *stamen*, warp, thread. See anther.

steep To extract flavor or medicinal qualities by soaking material in hot but not boiling liquid. See tea. Anglo-Saxon *steap*, bowl.

stipe A short supporting stalk; the leaf stalk of a fern, the stalk of a gill fungus. L. *stipes*, a tree trunk.

stipules Leaflike structures growing from the leaf base. Small "leaves" below flowers are bracts, those below leaves, stipules. L. *stipula*, stalk, diminutive of *stipes*, trunk.

stolon A slender, above-ground, horizontal stem which develops roots at node and/or tip; a runner. L. *stolo*, a shoot.

stoma, stomata, pl. Tiny openings through which leaves exchange gasses. Comparable openings in tree bark are called lenticels. Gr. *stoma*, mouth or pore.

strobilus, strobile Several modified leaves or ovule-bearing scales grouped together, usually conelike in shape; an ament. Gr. *strobilus*, a cone.

styptic An astringent. Gr. *stiptikos*, to constrict.

synonym A scientific plant name no longer in current use. Gr. *synonymon*.

tannin/tannic acid A plant substance with an astringent, bitter taste. A strongly astringent acid used in dyeing, tanning, medicine, etc. Indigestible to animals, it breaks down to protect plants from infection. French *tanin*, tanning principle.

taproot A main root growing downward. Scandinavian *rot*.

taxon/ taxa Classification units of related organisms, in ascending order: species, genus, family, order, class, phylum, kingdom. Gr.*taxis*, arrange.

taxonomy Science of classification dealing with arrangement of animals and plants according to their natural relationships. Gr.*taxis*, *nomos*, laws.

tea A solution prepared by pouring boiling water over plant parts, usually flowers, leaves, or small seeds, and allowing the material to steep; an infusion. From a Chinese dialect for tea.

tendril An organ by which plants climb and attach to a support. L. *tendere*, to stretch out, to extend.

thallus A plant body without true roots, stems, or leaves; characteristic thallophytes are algae, fungi, and lichens. Gr. *thallos*, a green shoot, from *thallein*, to sprout.

tincture A medicinal solution made with alcohol—sometimes because plant parts are not water soluble. L. *tingere*, to dye, stain, wet.

tree A perennial woody plant with a single stem or trunk, commonly exceeding ten feet in height. Anglo-Saxon *treo*, *treow*, tree, wood.

tuber The enlarged, short, fleshy, part of a rhizome, bearing "eyes" or buds; for example, a potato. L. *tuber*, a bump, a swelling.

umbel Flat or convex inflorescence with pedicels arising from a common point. L. *umbella*, sunshade.

vascular tissue Conducting tissue of plants. Xylem, (Gr. *xylon*, wood), the woody supporting inner part, conducts water and minerals; phloem, (Gr. *phloos*, bark), conducts foods. Cambium, when present, is between phloem and xylem. L. *vasculum*, a small vessel.

vermifuge A substance that destroys or expels intestinal worms; an anthelmintic. L. *vermis*, worm, *fugare*, to cause to flee.

virtues Active qualities or powers; early herbalists referred to "virtues" of medicinal plants. L. *virtus*, strength, courage, virtue, from *vir*, a man.

vulnerary Substance used to treat wounds. L. *vulnerare*, to wound.

weed A plant not valued for use or beauty, growing wild and rank and regarded as hindering growth of more desirable vegetation; any plant growing where it is not wanted. Anglo-Saxon *weod*, used in present meaning.

whorl A circle of leaves or flowers around the stem of a plant; probably a variety of whirl. Middle English *whirlen*, to whirl.

COLOPHON

There is a joy in volunteering. It is rewarding to see accomplishments made possible in part through your help. I enjoy reaching the end of a nature tour with a small hand in each of mine. One day I concluded a tour with the suggestion that guiding was a pleasure, and that perhaps the children or their parents might consider helping in the park.

A little second-grader looked up and said 'When I grow up, I want to do what you are doing!' If I have shared my love for the out of doors, of growing things, and my respect for our environment, I believe my influence will continue to live. This is my hope.

INDEX of PLANTS
Scientific names in italics
Italic numbers used for illustrations

Abies grandis, 87
Acer circinatum, 170, *171*
Acer macrophyllum, 114, *115*
Achillea millefolium, 182, *183*
Achlys triphylla, 168, *169*
Actaea rubra, 34
Adenocaulon bicolor, 128, 129
Adiantum pedatum, 113, 113
Alder, *30,* 31, 49, 102, 109, 148
Alliaria officinalis, 85, 85
Anaphalis margaritacea, 129, *130*
Anemone deltoidea, 181
Anthemis cotula, 68
Apocynum androsaemifolium, 67
Aquilegia formosa, 58
Arbutus menziesii, 112
Arctium minus, 44
Arrow Wood, 123, *124*
Aruncus sylvester, 86
Asarum caudatum, 177, 177
Ash, *12,* 32, 118
Aster, 33
Athyrium filix-femina, 104, *105*
Atropa belladonna, 37
Avens, *33, 33*
Baneberry, *34,* 44
Bedstraw, 28, 34, *35, 199*
Bellis perennis, 62, 187
Berberis spp. *2, 24,* 27, *124,* 125
Bigleaf Maple, 32, 102, 109, 114, *115,* 159, 185
Bindweed, 117
Bishop's Cap, *6,* 36
Bitter Cress, 36
Bittersweet Nightshade, 10, 29, 37, *37*
Blackberry, 38, *39*
Blackcap, 40, 164
Bladder Campion, 40
Blechnum spicant, 65, 66
Bleeding Heart, *40,* 41
Bog Orchid, 41
Brake Fern, 28, 41, *42,* 48, 70, 157
Brassica campestris, 119
Brooklime, 154, *154*
Buckbrush, 43
Buckthorn, 48, *48*
Bugbane, 44
Bull Thistle, 164
Bunchberry, 47
Burdock, 44
Burning Bush, 175

Buttercups, *45,* 46
Butterfly Leaf, 168, *169*
California Poppy, 47
Calypso bulbosa, 106
Canada Dogwood, 26, 47
Canada Thistle, 25, 164
Candy Flower, *20,* 25, 47
Capsella bursa pastoris, 150
Cardamine spp. 25, 151
Carex spp. 147
Cascara Sagrada, 48, *48*
Cat-tail, 50, 60
Cats-ear, 49, *49, 50*
Ceanothus sanguineus, 43
Cedar, 50, *51,* 71, 101
Chenopodium album, 106, 172
Cherry, 19, 52, 53, 123
Cherry Laurel, 106
Chickweed, 53
Chicory, 54
China Berry, 34
Chittim, 48
Cichorium intybus, 54
Cimicifuga elata, 44
Circaea alpina, 76, 76
Cirsium spp. 25, 164
Claytonia spp. 116
Cleavers, 34, *35*
Clematis ligustifolia, 54, *55*
Clover Broomrape, 55
Clover, 56
Coast Pine, 132, 147
Coltsfoot, 56, *57*
Columbine, 58
Coolwort, *28,* 28, 82
Compass Plant, 108, *108*
Conium maculatum, 136
Convulvulus arvensis, 117
Conyza canadensis, 63
Corallorhiza maculata, 58
Coral-root, 58
Corn lily, 78
Cornus canadensis, 26, 47
Cornus nuttallii, 27, 69, *202*
Cornus stolonifera, 68
Corylus cornuta, 91, 92
Cottonwood, 58, *59*
Cow Parsnip, 60
Crabapple, 60, *184*
Crane's Bill, 86, *86*

Crataegus spp. 90, *91*
Creeping Charlie, *61*, 62
Crepis spp. 90
Currant, 62
Cyperaceae, 147
Cytisus scoparius, 145, *146*
Daisy fleabane, 63
Dandelion, 27, 49, 54, 63, *64*, 89, 185
Daucus carota, 137, *138*
Dead-nettle, 66, 94
Deadly Nightshade, 37
Deer Fern, 65, 66
Deertongue, 80
Dewberry, 38
Dicentra formosa, 40, 41
Digitalis purpurea, 29, 68, *83*, 84, 146
Dipsacus sylvestris, 162
Disporum spp. 77, 78
Dock, 64, 66, 67, 155, 172
Dogbane, 67
Dogfennel, 68
Dogtooth Violet, 80
Dogwood, Creek, 68
Dogwood, 27, 69, *202*
Douglas Fir, 70, 101, 133
Dryopteris austriaca, *180*, 181
Duckfoot, *15*, 72
Eburophyton
Eburophyton austiniae, 131
Edible Thistle, 164
Elderberry, 9, 25, 31, 34, 74, 75, 109, 157, 164
Enchanter's Nightshade, 76, *76*
Endive, 54
English Daisy, 62, *187*
English Plantain, 134, *135*
Epilobium spp. 81, *82*
Equisetum spp. 27, 98, *98*, 147
Erodium cicutarium, 81
Erythronium spp. 80
Eschscholzia californica, 47
Euonymus occidentalis, 175
Evening Primrose, 76
Evergreen Violet, 25, 171
Fairy Bell, 77, 78
Fairy Lantern, 78
Fairy Slipper, 106
Fall Dandelion, 88
False Dandelion, 49
False Hellebore, 25, 78, *79*
False Mitrewort, *28*, 82
False Solomon's Seal, 25, 80
False Sol. Seal, Star-Flrd. *14*, 80
Fawn Lily, 80
Figwort, California, 81

Filaree, 81
Fireweed, 81, *82* .
Fir, 70, 87
Five-finger Fern, 113, *113*
Florida Dogwood, 70
Flowering Currant, 62
Foamflower *28*, 28, 82
Foxglove, 29, *83*, 84, 119, 146, 179, 185
Fragaria spp. 157, *158*
Fraxinus latifolia, 32
Fringecup, *83*, 84
Galium spp. 28, 34, *35*, *199*
Garlic Mustard, 29, 85, *85*
Gaultheria shallon, 142, *143*
Geranium spp. 86, *86*
Geum macrophyllum, *33*, 33
Glecoma hederaceae, 61, *62*
Glycyrrhiza glabra, 96, 109, *110*
Goatsbeard, 86
Goldenrod, 87
Gorse, 147
Goodyera oblongifolia, 139
Grand Fir, 87
Ground Ivy, 61, *62*
Groundsel, 88
Grove Lover, 120, *120*
Habenaria dilatata, 41
Hardhack, 89
Hawkbit, 89
Hawksbeard, 90
Hawthorn, 90, *91*
Hazelnut, *91*, 92
Heal-All, *93*, 94
Hedera helix, 102, *103*
Hedge Nettle, *93*, 94
Helleborus niger, 25
Hemlock, *95*, 96
Heracleum lanatum, 60
Heuchera spp. 32
Holly, 97
Holodiscus discolor, 123, *124*
Honesty, 97
Honeysuckle, 97, *186*
Horsetail, 21, 25, *26*, 98, 147
Horseweed, 63
Huckleberry, 99
Hydrophyllum tenuipes, 172, *173*
Hypericum perforatum, 142
Hypochaeris spp. 49, *49*
Ilex aquifolium, 97
Indian Pipe, 101
Indian Plum, *100*, 101
Inside-out-flower, *15*, 72
Iris spp. 102
Ivy, 102, *103*,

Klamath Weed, 142
Knotweed, *103*, 104
Lactuca spp, 108, *108*, 109
Lady Fern, 104, *105*
Lady's Slipper, 106
Lamb's Quarters, 106
Lambstongue, 80
Lamium purpureum, 66,
Lapsana communis, 121
Lathyrus latifolius, *128*, 129
Laurel cherry, 106
Lemon Balm, *11*, 27, 108
Leontodon spp. 89
Lettuce, *108*, 108, 109, *188*
Lichens, 17
Licorice Fern, 96, 109, *110*
Lily of the Valley, 179
Linnaea borealis, 167
Lodgepole Pine, 132
Lonicera ciliosa 97
Lonicera involucrata, 97, 167
Lunaria annua, 97
Lupine, 111
Lysichitum americanum, 29, 75, 133, 142, 150, 164
Madrona, 112, *112*
Mahonia spp. 27, 125, 126
Maianthemum dilatatum, *178*, 178
Maidenhair Fern, 113, *113*,
Maple, 114, 170
Marah oreganus, *176*, 176
Masterwort, 60
Matricaria matricarioides, 134
Mayweed, 68
Meadowrue, 116
Melissa officinalis, 27, 108
Mimulus spp. 117
Miner's Lettuce, *20*, 25, 47, 116
Mitella caulescens, 36
Mitrewort, 36
Mock Orange, 73, 116
Money Plant, 97
Monkey flower, 117, *117*
Monotropa uniflora, 101
Montia perfoliata, 116
Montia sibirica, 20, 25, 47
Morning Glory, 117
Mosses, 21
Mountain Ash, *23*, 33, 118
Mullein, 119
Mustard, 119
Nemophila parviflora, *120*, 120
Nettle, 27, 66, 94, 155
Nightshade, 37
Ninebark, 121, *121*
Nipplewort, 121
Nootka Rose, 139

Oak, 29, 109, 122
Ocean Spray, 53, 123, *124*
Oemleria / Osmaronia cerasiformis, *100*, 101
Oenanthe sarmentosa, 174, *175*
Oenothera spp. 76
Orchids, 18
Oregon Grape, *2*, 24, 27, 72, *124*, 125
Oregon Tea, 43
Orobanche minor, 55
Osmorhiza chilensis, 158, *159*
Oxalis spp. 126, *127*
Oxeye Daisy, 127, *128*
Oyster Plant, 145
Pathfinder, *128*, 129
Pearly Everlasting, 129, *130*
Perennial Pea, 29, *128*, 129, 170
Petasites spp. 56, 57
Phacelia nemoralis, *173*, 174
Phantom Orchid, 131
Philadelphus lewisii, 116
Physocarpus capitatus, *121*, 121
Piggy-Back Plant, 131, *131*
Pigweed, 106, 148, 172
Pine, 71, 132, 147
Pineapple Weed, 134
Pinus spp. 71, 132, 147
Plantago spp. 29, 134, *135*, 157
Plantain, 29, 134, 157
Poison Hemlock, 136
Poison Ivy, Oak, 136
Polygonum aviculare, *103*, 104
Polypodium spp.109, *110*
Polystichum munitum, 159, *160*
Ponderosa Pine, 71, 131
Populus trichocarpa, 58, *59*,
Prickly Lettuce, 108, *108*
Prunella vulgaris, *93*, 94
Prunus emarginata, 19, 52, 53, 123
Prunus laurocerasis, 106
Pseudotsuga menziesii, 79
Pteridium aquilinum, 41, 42
Purple Violet, 171
Pyrus fusca, 60
Queen Anne's Lace, *137*, 138
Quercus garryana, 122
Ragwort, 88
Ranunculus spp. 45, 46
Rattlesnake Plantain, 139
Redwood Sorrel, 126
Rein Orchid, 44
Rhamnus spp. 48
Rhapanus sativus, 139
Rhus spp. 136
Ribes sanguineum, 62
Rosa spp. 139-141
Roses, 139-141
Rowan, 118

Rubus spp. 38-40, 39
Rubus parviflorus, 163, 163
Rubus spectabilis, 143, 144
Rumex spp. 66, 67, 149, 149
St. Johnswort, 25, 142
Salal, 29, 30, 60, 75, 125, 142, 143
Salix spp. 179
Salmonberry, 29, 143, 144, 163
Salsify, 145
Sambucus spp.72-75, 74
Saskatoon Berry, 148, 148
Scirpus spp. 147
Scotch Broom, 145, 146
Scouring Rush, 147
Scrophularia californica, 81
Sedges, 147
Senecio jacobaea, 161, 161
Senecio vulgaris, 88
Service Berry, 29, 148, 148
Shady Waterleaf, 173, 174
Sheep Sorrel, 66, 149, 149
Shepherd's Purse, 150
Siberian Miner's Lettuce, 20, 25, 47
Silene cucubalis, 40
Silver Dollar, 97
Silvergreen, 128, 129
Skunk Cabbage, 29, 75, 133, 142, 150, 164
Small Bittersweet, 10, 29, 37, 37
Smilacina spp. 80
Snowberry, 151, 152
Solanum dulcamara, 10, 29, 37, 37
Solidago spp. 87
Solomon's Seal, 25, 80
Sonchus spp. 153, 153
Sorbus spp. 118
Sorrel, Sheep, 149, 149
Sorrel, Redwood, 126, 127
Sourgrass, 66 149, 149
Sow Thistle, 153, 153
Speedwell, American, 154, 154
Spiraea douglasii, 89
Spring Beauty, 25, 47, 151
Stachys cooleyae, 93, 94
Starflower, 154, 154
Steeple bush, 89
Stellaria media, 154
Stork's Bill, 81
Stinging Nettle, 27, 64, 67, 73, 76, 88, 94, 149, 155, 155
Strawberry, 157, 158
Streptopus spp. 168
Sweet-after-death, 168, 169
Sweetbriar Rose, 141
Sweet Cicely, 26, 158, 159
Sword Fern, 159, 160
Symphoricarpos albus, 151, 152
Syringa, 73, 116
Tanacetum vulgare, 162
Tansy Ragwort, 29, 88, 161, 161
Tansy, 162

Taraxacum officinale, 27, 63, 64
Taxus brevifolia, 183
Teasel, 162
Tellima grandiflora, 83, 84
Thalictrum occidentale, 116
Thimbleberry, 29, 157, 158, 163, 163
Thistle, 25, 164
Thuja plicata, 50, 51
Tiarella trifoliata, 28, 28, 82
Tiger Lily, 165
Tolmiea menziesii, 131
Toothwort, 25, 151
Tragopogon spp. 145
Trientalis latifolia, 154, 154
Trifolium spp. 56
Trillium ovatum, 165, 166, 185
Tsuga heterophylla, 95, 96
Twin-berry, 8, 167
Twinflower, 167
Twisted Stalk, 168
Typha latifolia, 50, 60
Ulex europaeus, 147
Urtica dioica, 27, 64, 67, 73, 76, 88, 94, 149, 155, 155
Vaccinium spp. 99, 100
Vancouveria hexandra, 15, 72
Vanilla leaf, 168, 169
Veratrum spp. 78
Verbascum thapsus, 119
Veronica americana, 154
Vetch, 129, 169, 170
Vicia spp. 129, 169, 170
Vine Maple, 170, 171
Viola spp. 18, 171, 172
Wake Robin, 165
Waterleaf, 172, 173
Water Parsley, 174, 175
Waxberry, 151, 152
Western Wahoo, 175, 195
Wild Carrot, 137, 138
Wild Cucumber, 176, 176
Wild Ginger, 25, 177, 177
Wild Lily of the Valley, 178, 178
Wild Pea, 129
Wild Radish, 139
Willow Herb, 81, 82
Willows, 179
Windflower, 181
Wood Fern, 104, 180, 181
Wood Rose, 141, 141
Wood Sorrel, 126, 127
Wood Violet, 25, 171, 172
Woodland Star, 154, 154
Woods Buttercup, 45, 46
Woody Nightshade, 10, 29, 37, 37
Woundwort, 93, 94
Yarrow, 182, 183
Yew, 183
Yellow Avens, 33, 33
Youth on Age, 131, 131